RISK-BASED DECISIONMAKING IN WATER RESOURCES X

PROCEEDINGS OF THE TENTH CONFERENCE

November 3–8, 2002
Santa Barbara, California

SPONSORED BY
United Engineering Foundation

CO-SPONSORED BY
Universities Council on Water Resources

Water and Environmental Planning and Management Committee
Environmental and Water Resources Institute (EWRI) of the American Society of Civil Engineers (ASCE)

SUPPORTED BY
National Science Foundation

US Army Corps of Engineers, Institute for Water Resources

EDITED BY
Yacov Y. Haimes
David A. Moser
Eugene Z. Stakhiv

TECHNICAL EDITOR
Grace Ivry Zisk

TECHNICAL ASSISTANTS
Della Dirickson
Burton I. Zisk

Published by American Society of Civil Engineers

Library of Congress Cataloging-in-Publication Data

United Engineering Foundation Conference on Risk-Based Decisionmaking in Water Resources (10th : 2002 : Santa Barbara, Calif.)
  Risk-based decisionmaking in water resources X : proceedings of the tenth conference, November 3-8, 2002, Santa Barbara, California / edited by Yacov Y. Haimes, David Moser, Eugene Z. Stakhiv ; sponsored by United Engineering Foundation ... [et al.].
    p. cm.
  Includes bibliographical references and index.
  ISBN 0-7844-0694-4
  1. Water resources development--Decision making--Congresses. 2. Risk assessment--Congresses. 3. Water-supply--Security measures--United States--Congresses. 4. Terrorism--United States--Prevention--Congresses. I. Title: Risk-based decisionmaking in water resources X. II. Haimes, Yacov Y. III. Moser, David A. IV. Stakhiv, Eugene Z. V. United Engineering Foundation (U.S.) VI. Title.

TC401.U5354 2002
363.6'1--dc22   2003057744

American Society of Civil Engineers
1801 Alexander Bell Drive
Reston, Virginia, 20191-4400

www.pubs.asce.org

Any statements expressed in these materials are those of the individual authors and do not necessarily represent the views of ASCE, which takes no responsibility for any statement made herein. No reference made in this publication to any specific method, product, process, or service constitutes or implies an endorsement, recommendation, or warranty thereof by ASCE. The materials are for general information only and do not represent a standard of ASCE, nor are they intended as a reference in purchase specifications, contracts, regulations, statutes, or any other legal document. ASCE makes no representation or warranty of any kind, whether express or implied, concerning the accuracy, completeness, suitability, or utility of any information, apparatus, product, or process discussed in this publication, and assumes no liability therefore. This information should not be used without first securing competent advice with respect to its suitability for any general or specific application. Anyone utilizing this information assumes all liability arising from such use, including but not limited to infringement of any patent or patents.

ASCE and American Society of Civil Engineers—Registered in U.S. Patent and Trademark Office.

Photocopies: Authorization to photocopy material for internal or personal use under circumstances not falling within the fair use provisions of the Copyright Act is granted by ASCE to libraries and other users registered with the Copyright Clearance Center (CCC) Transactional Reporting Service, provided that the base fee of $18.00 per article is paid directly to CCC, 222 Rosewood Drive, Danvers, MA 01923. The identification for ASCE Books is 0-7844-0694-4/03/ $18.00. Requests for special permission or bulk copying should be addressed to Permissions & Copyright Dept., ASCE.

Copyright © 2003 by the American Society of Civil Engineers.
All Rights Reserved.
Library of Congress Catalog Card No: 2003057744
ISBN 0-7844-0694-4
Manufactured in the United States of America.

Preface

The first United Engineering Foundation Conference on Risk-Based Decisionmaking in Water Resources was held at the Asilomar Conference Grounds, Pacific Grove, California on September 21-26, 1980. Twenty-two years later, the tenth conference assembled on November 3-8, 2002 in Santa Barbara, California. Although significant social, political, and technological changes have taken place during the last two decades, many of the issues that were on the agenda in 1980 remained relevant in 2002. The major exception was a stronger focus on terrorism at the tenth conference, although this subject was also on the agenda of the eighth and ninth conferences.

These proceedings of the Tenth United Engineering Foundation Conference on Risk-Based Decisionmaking in Water Resources illustrate that our original objectives and goals, as well as the issues raised, remain relevant, important, and timely. Influenced by the terrorist attacks on the US on September 11, 2001, the theme of the tenth conference was *Risk of Terrorism*. Indeed, the entire first day was devoted to the risk and vulnerabilities to terrorism of the homeland's water resources system of systems. The second day addressed the same topic from the methodological perspectives of risk of extreme events. The third day examined the interconnectedness and interdependencies between the water resource system of systems and other infrastructures. The sessions on the fourth day focused on institutional and organizational structuring associated with risks of terrorism. As always, the session on the last day (Friday morning) was devoted to reflections on the participants' responses to the questionnaires. Recognizing that these are the tenth such proceedings, we have compiled the responses to the same three questions from all ten conferences and included them at the end of this book. We have begun analyzing the trends that emerged, and the results should make an important contribution to the field of risk-based decisionmaking in water resources.

Following is a brief summation of the major themes of past conferences:

The second conference, in November 1985, expanded on the 1980 theme. It was "geared toward balancing issues of practical concern as well as understanding some of the philosophical underpinnings and theoretical premises of risk analysis."

In the preface to the third conference, held in November 1987, we asked: "What has changed during the past decade to warrant yet another conference?" To answer this question, we stated that the conference's goals and objectives were to:
1) familiarize the participants with the state of the art in risk/benefit analysis,
2) explore the feasibility of using risk/benefit analysis in water resources planning and management,
3) provide a medium conducive to the exchange of information on the conference theme among educators, analysts, managers, and policy makers, and
4) identify and articulate future desired actions designed to alleviate some of the present problems we face in risk/benefit analysis and risk assessment in general.

What had changed in 1987 was that in this still-new professional niche, we had matured.

During the fourth conference in October 1989, we searched for the driving forces leading to the growing popularity of risk-based decisionmaking. We attributed the increasing popularity and prominence of risk analysis to two basic factors: society and technology.

At the opening of the fifth conference in November 1991, we quoted a chief executive officer who told his board of directors: "Our role is to manage change; if we can't manage change, we must change management." Noting that "the field of risk analysis is changing fast and we must not be left behind," we stated:

- We meet in these conferences to exchange information and knowledge on the changes that are taking place in the field.
- We must facilitate the process of listening to each other in these meetings by getting to know each other on a personal basis.
- We must create an environment that is conducive to dialogue and communications: fewer lectures and more discussions.
- We must open new lines of communication.
- We must be able to challenge ourselves and let others challenge our old assumptions.
- We must free ourselves of prejudices.
- We must be ready to reexamine our biases—professional, personal, and other biases.
- We must be able to learn from each other, to discover what new theories and methodologies have been developed and where they have been applied—either successfully or with less success.
- We should be ready to accept the premise that risk management must be an integral part of a total systems management, and adopt a holistic philosophy.

The sixth conference, held in 1993, continued to reinforce the Socratic culture that had evolved in these meetings. Although some of the papers covered topics presented previously, the discussions were more substantive and in greater depth. Methodologies were more closely related to theory, and at the same time the relevance of their applications to emerging natural and man-made hazards became stronger and more convincing. Such topics as uncertainties in data, models, and forecasts and their influences on risk analysis have, in some sense, an eternal life of their own; yet the level of discussion epitomized the growth and maturity in the field.

The seventh conference, in October 1995, augmented the technical discussion with policy issues and the implications of recent legislative initiatives in risk assessment. It attempted to address the connectedness among such emerging trends and ideas as the management of our environment, physical infrastructure, response to possible climate change, the desire to embrace the concept of sustainable development in its broader sense, and the explosion of communications opportunities and their impacts on informed decisionmaking.

The eighth conference, held in October 1997, continued exploring these themes, and also focused on the climatic effects of El Niño.

In the ninth conference, in October 2000, we continued exploring these themes and reviewed our approaches to risk and uncertainty during the past 20 years. We addressed the economic dimensions of risk analysis, continued the search for answers to the survivability of our critical infrastructures under natural and willful threats, and explored new approaches to risk of extreme and rare events. In 2000, at the start of a new millennium, those of us involved in risk-based decisionmaking continued to experience the same evolutionary process that systems analysts and systems engineers went through earlier. However, as in 1987, there remained many who saw risk analysis as simply a specialized extension of the body of knowledge and evaluation perspectives that had come to be associated with systems analysis.

In 2000 and again in 2002, we were even more certain that risk assessment and management must be an integral part of the decisionmaking process, rather than a gratuitous add-on technical analysis. Some of us are becoming more and more convinced of the grave limitations of the traditional and commonly-used expected value concept, and are complementing and supplementing this concept with conditional expectations, where decisions about extreme and catastrophic events are not averaged out with more commonly occurring high-frequency/low-consequence events.

Today, more than two decades after the first conference in 1980 in Asilomar, California, there is a strong public awareness of the subject of risk: environmental risks, technological and natural risks, human health and safety risks, and risks to our critical infrastructure and to cyberspace. The professional community is responding much more forcefully and knowledgeably as well, and in many instances, leading what has ultimately come to be a political debate. We are more critical of the tools that we have developed, because we recognize their ultimate importance and usefulness in the resolution of critical societal problems. We are more willing to accept the premise that in most cases, a truly effective risk analysis study must be cross-disciplinary, relying on social and behavioral scientists, engineers, regulators, and lawyers.

These are some of the trends that have distinguished the conferences since 1980, and continue today. We hope that the interest in these United Engineering Foundation conferences remains as high as it has been in the past and we look forward to the eleventh conference in 2004.

Several organizations and individuals were instrumental in making this conference possible. Thanks to Dr. Douglas James of the National Science Foundation for his support of this conference. Bob Pietrowsky, Director of the US Army Corps of Engineers Institute for Water Resources, and Jerry Foster, Engineering and Construction Division, Headquarters, US Army Corps of Engineers, Washington, DC, were generous in providing financial support for the conference, as was Dr. Charles Freiman of the United Engineering Foundation. We also thank Richard Fein, the

Engineering Foundation Conference Liaison, and Antoinette Chartier, the Engineering Foundation Conference Local Coordinator.

In addition, we thank the ASCE Task Committee on Risk-Based Decisionmaking and the Universities Council on Water Resources for again co-sponsoring this conference and enabling a large number of prominent speakers and participants to present and exchange ideas. Ultimately, the value of such a conference lies in the influence that these ideas and their presenters have on the vital issues and frequently extraordinary events occurring in our rapidly changing world.

All papers have been reviewed, edited, and accepted for publication in these proceedings by the editors. The papers are eligible for discussion in the *Journal of Water Resources Planning and Management* and also eligible for ASCE awards.

Finally, we acknowledge the invaluable editorial and computer work provided by Grace and Burton Zisk, the significant contributions of our colleague James H. Lambert to the organization of the conference, the administrative assistance provided by Della Dirickson, Manager, Center for Risk Management of Engineering Systems, University of Virginia, Barbara Hickernell and the staff of the United Engineering Foundation, as well as Donna Dickert and the ASCE staff for their hard work in bringing these proceedings to their final printed form. Thanks also to Joost Santos, Ruth Dicdican, and Kenneth Crowther for their contributions in compiling the responses to the questionnaire.

                                      Yacov Y. Haimes
                                      Charlottesville, Virginia

                                      David A. Moser and Eugene Z. Stakhiv
                                      Fort Belvoir, Virginia

Contents

Preface iii
    Yacov Y. Haimes, David A. Moser, and Eugene Z. Stakhiv

Risk Analysis in Disaster Planning by Superposition of Infrastructure and Societal Networks 1
    James H. Lambert and Priya Sarda

The Fuzzy Logic Paradigm of Risk Analysis 12
    Istvan Bogardi and Lucien Duckstein

A Dynamic Risk Model for Information Technology Security in a Critical Infrastructure Environment 23
    John H. Saunders

Quantifying and Communicating Model Uncertainty for Decisionmaking in the Everglades 40
    Daniel P. Loucks

Optimal Allocation of Resources for Defense of Simple Series and Parallel Systems from Determined Adversaries 59
    Vicki M. Bier and Vinod Abhichandani

Applying the General Theory of Quantitative Risk Assessment (QRA) to Terrorism Risk 77
    Stan Kaplan

Vulnerability of Water Systems to Acts of Terrorism and Acts of Nature 82
    Nicholas C. Matalas

Toward a Systems-Based Vulnerability Assessment Methodology for Water Supply Systems 91
    Barry C. Ezell

Demand-Reduction Input-Output (I-O) Analysis for Modeling Interconnectedness 104
    Joost R. Santos and Yacov Y. Haimes

Belief Systems and Reducing Risks from Terrorism 119
    Robert E. O'Connor

GIS Model for Estimating Dam Failure Life Loss 126
    Maged Aboelata, David S. Bowles, and Duane M. McClelland

Disintegrated Water Resources Management in the US: The Union of Sisyphus and Pandora 146
    Eugene Z. Stakhiv

Vulnerability to Terrorism: Addressing the Human Variables 155
    William D. Rowe

Session 3 Summary—Lessons Learned from Experience Dealing with Risks of Extreme Events: Part I 160
    Ruth Y. Dicdican

Session 4 Summary—Lessons Learned from Experience Dealing with Risks of Extreme Events: Part II 163
    Joost Santos

Session 8 Summary—Panel Discussion: Synthesis—What Does It All Mean? 168
    Jim Lambert

Summary of Responses to Participant Questionnaires 171

List of Participants 217

Program 219

Subject Index 227

Author Index 229

Risk Analysis in Disaster Planning by Superposition of Infrastructure and Societal Networks

James H. Lambert[1] and Priya Sarda[2]

Abstract

This paper addresses disaster planning for infrastructures and their related societal entities through innovative modeling that coordinates multiple networks against the contingency scenarios that disrupt plans for mitigation, preparedness, response, and recovery. Across disciplines, interdependent infrastructures and related societal entities are increasingly amenable to characterization as mathematical networks. Effective disaster planning for complex systems is known to rely on the identification and management of a multitude of diverse contingency scenarios. Modeling of such contingencies in disaster planning for interdependent infrastructures and societal entities can proceed with a superposition of multiple network models. We study the interactions among contingency scenarios and diverse superposed networks. The developed theory and methodology will guide the multidisciplinary selection and management of relevant worst-case scenarios, an essential activity for disaster planning in large-scale, complex systems.

Introduction

Planning for large-scale disasters such as hurricanes, floods, earthquakes, and terrorism involves phenomena across interdependent transportation, telecommunications, water, power, and related societal entities such as religious and ethnic groups, health care, human services, and tourism. Amin [2002] identifies critical infrastructure systems as cable and wireless telecommunications; banking and finance; land, water, and air transportation; gas, water, and oil pipelines; electric

---

[1]Research Assistant Professor and Associate Director, Center for Risk Management of Engineering Systems (CRMES) and Department of Systems and Information Engineering (SIE), University of Virginia, Charlottesville, VA 22903; 434-982-2072; lambert@virginia.edu
[2]Graduate student, CRMES-SIE, University of Virginia, Charlottesville, VA 22903; 434-982-2072; ps2y@virginia.edu.

power grids; and the internet, which can be international, national, state, and local in scope. Haimes and Jiang [2001] describe a need to assess the vulnerabilities of complex interdependent infrastructures. Hecker et al. [2000], Ezell and Farr [2000], and Heaney et al. [2000] further address the risk of disasters to infrastructures. Juhl [1993] describes efforts of FEMA to increase the efficiency of pre- and post-disaster plans by GIS database applications. Ardekani [1992] evaluates the response following the 1989 Loma Prieta earthquake and provides recommendations for the preparedness of transportation agencies in the future. Kovel [2000] addresses modeling of disaster response. Mondul [1997] describes information-sharing among agencies relying on the transportation system in the preparation and recovery processes. Parentela et al. [2000] describe a variety of disaster-planning factors, including the environment, the capabilities of emergency response providers, and the economy. Disaster planning is further addressed by practitioners including ASIS International [2003], ASIS [2003], Myers [1999], Erickson [1999], and Gigliotti and Jason [1991]. Haimes et al. [2002], Lambert et al. [2001], and Haimes [1998] perform risk identification for system hierarchies, using hierarchical holographic modeling [Haimes 1981; Hall 1989] to effect a superposition of hierarchical systems. Sharit [2000] uses hierarchies of factors for risk identification. Lambert and Patterson [2002] develop risk identification for delay scenarios in the hurricane recovery of a transportation agency. Kaplan and Garrick [1981] provide a framework for risk analysis. Wei [1991] describes failure mode, effects, and criticality analysis in risk analysis. Watts [2003], Barabási [2002], and Arquilla and Ronfeldt [2001] address the impetus to model various infrastructure and societal systems as networks and, for an example, Marburger and Westfechtel [2002] study telecommunications networks. Buckley and Lewinter [2002] and Xu [2001] describe advances in the foundations of graph theory.

We recognize an opportunity to better understand disaster planning that surrounds infrastructures and societal entities. Toward such ends, mathematical networks are suited to the study of advanced relationships among contingency scenarios in these areas. Specifically, we develop an approach to improve multidisciplinary disaster planning through the integration of risk identification with multiple network analyses. Interdependent networks of both critical infrastructures and the related societal entities are addressed.

The organization of this paper is as follows. The methodology section describes the relationship of contingency scenarios to interdependent networks. The example section describes an application of the methodology to contingencies involving three networks. Finally, the conclusions describe the context for application of the methodology in the refinement of disaster planning, including preparedness and mitigation, response, and recovery.

Methodology

Following Lambert and Sarda [2003] and Sarda [2003], the following describes infrastructure systems and related societal organizations as networks for correspondence to a set of contingency scenarios that can be disruptive to disaster plans.

Let $G = \{ G_i \}_i$

denote the set of infrastructure and related societal networks. The networks are indexed by the set i = A to Z, AA to ZZ, etc. The ith network is described by

$G_i = (N_i, A_i)$, where

i ε {A,B,......,Z, AA....ZZ....etc.}

For example, Network "A" can represent an oil pipeline system, Network "C" can represent a cultural, political, or religious network, Network "M" can represent a telecommunication network, etc. Furthermore, let N denote the set of all nodes in the superposition of the multiple networks as follows:

$N = \{ N_i \}_i$

where N_i is the set of nodes of the ith network.

$N_i = \{ N_i^k \}_k$

where k ε positive integers indexing the nodes of a Network "i." Similarly, let A denote the set of all arcs. Each individual arc, $A_i^{l,m} = (N_i^l, N_i^m)$, is an ordered pair of two nodes as shown. The order indicates the direction of the arc in the network diagram.

$A = \{ A_i \}_i$ where $A_i = \{ (N_i^l, N_i^m) \}_{l,m}$

where l,m ε positive integers indexing the nodes of Network "i."

Let U denote the set of contingency scenarios that are developed for disaster-planning trouble-shooting:.

$U = \{S_1............S_j\}$, where

j ε positive integers and j ε {1,.........,J}.

Let S_0 denote the as-planned scenario involving disaster mitigation, preparedness, response, and/or recovery activities. A sample of contingency scenarios can be compiled by using historical data, brainstorming with cultural and religion experts, and conducting interviews with experts on the infrastructure and societal networks. A database of such activity is comprised minimally of the following data fields: description of the scenario, direct interactions (arcs and nodes) identified for each scenario, and indirect interactions (arcs and nodes). In turn, these are described as follows: an interaction arises when either singular or a combination (higher order) of network components is related to a contingency scenario.

Define a family of operators Θ_d to obtain network interactions that are directly involved in a contingency scenario. The operation that returns the arcs and nodes that are in direct interaction with contingency scenario "j" is

$\Theta_d(S_j)$.

The output will be in the form of nodes and arcs from across the superposition of networks G. To obtain only arc interactions directly involved in contingency scenario "j", the operation is modified to

$\Theta_{dA}(S_j)$.

For example, arc interactions that are directly involved with the scenario S_{0006}, disruption of information systems, are given by the operation

$\Theta_{dA}(S_{0006})$

which returns the arcs (C02,C05),(F05,F17) corresponding to a telecommunications link between two servers and the corridors of an office building. Similarly, to obtain only node interactions directly involved in contingency scenario "j", the operation is modified to

$\Theta_{dN}(S_j)$.

For example, node interactions that are directly involved with the scenario S_{0630}, contamination of a drinking water supply, are given by the operation

$\Theta_{dN}(S_{0630})$

which returns the nodes K02,K03,W08 corresponding to an elevated water tank, a water-pipe junction, and an e-business entity used for the purchase/logistics of chemicals.

Direct interactions are *prima facie* relationships of network components and contingency scenarios. Advanced relationships can be identified to describe rippling influences among network components. The study of advanced relationships can lead to understanding the subtle interconnectivity and interdependency across superpositioned networks. Thus, let q index the several definitions of rippling effects that propagate across nodes and arcs of the networks (to be explored in the research tasks).

Let Θ_q be an operator used to obtain the network components indirectly associated with a contingency scenario, and a particular relationship "q" is given by the operation

$$\Theta_q(\Theta_d(S_j)).$$

The operation has the set of direct node and arc interactions of a particular contingency scenario as its input, resulting in an output of nodes and arcs indirectly associated with the contingency scenario. For example, the node interactions that are indirectly associated with contingency scenario S_{0202}, an event that occurs on a cultural or religious holiday, is given by the operation

$$\Theta_q(\Theta_{dN}(S_{0202}))$$

which returns nodes for a particular rippling effect "q." This includes H01, J06 components which are objects of the rippling effects of the contingency scenario S_{0202}, and not direct interactions that are members of the set $\Theta_{dN}(S_{0202})$.

An interaction of a contingency scenario with networks can be singular or in pairs, triples, etc. A singular interaction is just a node or an arc. A paired interaction is a combination of components such as arc-arc, node-node, and arc-node. The higher-order interactions are studied to investigate interdependencies among the networks associated with a contingency scenario or collection of scenarios. The pairs of arcs interacting with a contingency scenario "j" is given by

$$\{(A_i^{l,m}, A_k^{l',m'}) : A_i^{l,m}, A_k^{l',m'} \; \varepsilon [\Theta_q(\Theta_d(S_j)) \cup \Theta_d(S_j)]$$

where l, m, l', m' are integers and when i = k, where (l,m) is not equal to (l',m').

That is, individual arcs $A_i^{l,m}$ and $A_k^{l',m'}$ are a subset of the direct and indirect components related to the jth scenario. For example, node pairs that are interacting with a contingency scenario TS_{0685}, shutdown of a nuclear reactor, is given by the operation

$$\{(N_i^l, N_k^s) : N_i^l, N_k^s \; \varepsilon \; [\Theta_q(\Theta_d(S_{0685})) \cup \Theta_d(S_{0685})] \}.$$

The node pair interactions are (V04,V12),(L56,L15), corresponding to a section of highway abutting the nuclear facility and a source of cooling water for the reactor.

Now define a family of operations to count the various nature of interactions. The cardinality, defined as η ( ), is a measure of number of interactions. For example η (N_B) is the number of node interactions in Network "B", a railway system. Similarly η (A_G) will count the number of arcs in Network "G", a distribution network of cereal grains. The cardinality of paired arc indirect interactions of contingency scenario S_{0890}, the release of panic-inducing information to the public, is given by the operation

$$\eta \{(A_i^{l,m}, A_k^{l',m'}) : A_i^{l,m}, A_k^{l',m'} \; \varepsilon \; [\Theta_q(\Theta_d(S_{0890})) \cup \Theta_d(S_{0890})] \}.$$

The cardinality of a particular interaction set being high indicates a need for further examination of the highlighted relationship between the superpositioned networks G and the contingency scenarios U.

Example

Consider a contingency scenario TS_{0005}, the unavailability of an oil pipeline section. The scenario can be related to the network representations of the following interdependent network systems: oil pipeline system, railway system, and societal organization. The modeled networks describe the physical characteristics of the infrastructure system by nodes and arcs. The nodes in Network "A" are pipeline junctions and the arcs connecting the nodes represent the pipeline sections. Similarly, in Network "B" the nodes are the railway intersections or station and the arcs are the connecting tracks. Similarly, in Network "C" the nodes represent the entities of the societal organization and the arcs are the information and physical distribution channels. The node label consists of the indexing alphabet of the particular network followed by the numeral representing the node position in the network. The notations as applied to the node labels are of the form "A3", depicting the junction at Santa Clara in the network representation of oil pipeline network "A". The arc labels are of the form (A4,A5), describing the oil pipeline section between junctions: San Jose (A4) and Evergreen (A5). The node interactions directly involved in the scenario S_{0005} are A4, A3, B6, B5, and C6. The node representations are the junctions of the affected pipeline sections: A4, San Jose; A3, Santa Clara; B6, the railway station at San Francisco; B5, the railway station at San Jose; and C6, an entity of the societal organization. The indirect interactions are defined by their relationships with the direct components and can span across multiple networks. The interactions that are indirectly associated with the contingency scenario S_{0005} are the nodes A1 and C5 and arc (B6,B7). A1 is the junction at Alviso, C5 is the organizational entity at Berlin, and the arc (B6,B7) is a track connecting the nodes B6, San Francisco, and B7, Matlock.

The cardinality of direct interactions involved with individual contingency scenarios is as follows: The contingency scenario S_{0390}, contamination in the food supply, has 48 direct node interactions and 60 direct arc interactions. The contingency scenario S_{0571}, loss of capability for water treatment, has only two direct node interactions and four direct arc interactions, indicating a close-knit societal organization operation in comparison with the magnitude of the scenario S_{0390}. The ripples of the contingency scenarios can span multiple infrastructure systems, supporting the claim that disjoint infrastructure systems are interdependent with respect to the execution of a disaster plan.

The infrastructure systems can be related to multiple scenarios to identify the system interfaces most in need of attention in disaster planning. Such relevant information, extracted from contingency scenario lists, consists of the infrastructure system and the different scenarios with which it has been involved. For example, the Network "E" electric power system has appeared in two scenarios, S_{0005} and S_{0010}.

The infrastructure network appearing most frequently on the scenario list should be of greater interest in disaster planning. For example, Network "O", the internet, is the most prominent infrastructure system across different contingency scenarios, suggesting a well-developed knowledge base of internet technology in the societal organization. Such analysis will alert agencies and officials to take further steps which might well involve reconfiguration of a disaster plan.

Frequency diagrams, which depict the cardinality of interactions of a contingency scenario, can be developed in the next step of the analysis by identifying the most prevalent interactions. These will be used in reconfiguring disaster planning against any threat. For example, we can consider the direct interactions across all contingency scenarios. The dominating component is the arc (C2,C5) which has approximately 340 interactions. The arc (C2,C5) represents the flow of information between the organizational cells C2, Boston, and C5, Berlin, indicating the most active component of the societal organization "C". The information compilation about principal interactions involved in a scenario is a useful precursor to risk modeling. Pair interactions are a result of interactions between the components of networks related to a contingency scenario. The different interactions are arc-node, node-node, and arc-arc interactions. An arc-node interaction, a variant in the pair interactions, is the node C6 related to the arc (B4,B5). This can be interpreted as the use of a section of the railway system "B" by a member of the organizational cell C6 in the scenario S_{0390}.

We can consider as well the greatest number of direct pair interactions. The paired components typically have fewer interactions than the dominant singular components. Few interactions in a paired interaction diagram mean that there is very little interdependency between the infrastructure networks in the context of a set of contingency scenarios. In an iteration of the analysis, the cardinality of the indirect interactions are calculated across the set of contingency scenarios.

Next, consider the cardinality of the indirect interactions across different contingency scenarios. Indirect relationships broaden and enhance the scope of investigation by studying the less-apparent interactions between network components. For example, node Z6 was not of particular significance in the direct interaction study. But the node Z6, a fundraising center in Network "Z", has approximately 150 interactions and is the most prominent indirect interaction. Arcs and nodes indirectly associated with involved arcs and nodes can help to identify the significant paths of warning or rescue operations. By securing the involved and indirectly associated arcs and nodes, risk of an attack could be mitigated.

Now consider the most prevalent paired interactions of indirect (rippling effects) components across all scenarios. A plot of interaction frequency can indicate a relationship between the pair of interactions and must be further investigated to explain that relationship. For example, the paired arc (C1,C4),(Z5,Z6) has approximately 30 interactions. The arc (C1,C4) represents the flow of information between the Atlanta and Denver cells of the societal organization. The arc (Z5,Z6) represents a financial network transaction between the United States and Africa. Such analysis proceeds iteratively in refining the disaster-planning scenario S_0, network models G, and the contingency scenarios U.

Conclusion

The implementation and context of our approach for characterizing contingency scenarios in multidisciplinary disaster planning are as follows: an initial set of contingency scenarios affecting disaster planning for infrastructure and societal networks is identified by planners and risk managers. Direct interactions of the scenarios and network components are identified across the superposition of networks. Indirect interactions of scenarios and network components are generated through study of the rippling of scenarios across network components. Rippling is characterized by proximity, reliability-cut-set methods, and shortest-path algorithms. Higher-order interactions are identified when multiple network components are associated, directly or indirectly, with the same contingency scenario. The relationships of the scenarios to the networks are thus studied by the combination of the several modes: (i) direct and indirect interactions, and (ii) singular and higher-order interactions. Frequency analyses of the interactions among scenarios and the networks will be useful to characterize the rippling effects of the scenarios across the superposition of networks. The approach supports iterative refinement of disaster-response plans based on the relevant contingency scenarios. The paper addresses a fundamental research direction at the multidisciplinary intersection of risk analysis and network systems analysis to provide understanding of and ultimately improve disaster planning for complex systems. Future work should address the implications of the analysis approach effort for operational-support databases in disaster planning.

Acknowledgments

The authors are grateful to the participants of the Engineering Foundation Conference on Risk-Based Decisionmaking in Water Resources X, Santa Barbara, CA, November 2002, for their comments and suggestions; to Yacov Y. Haimes, Barry M. Horowitz, Irwin M. Pikus, and Gregory B. Saathoff, all of the University of Virginia, for related discussion and collaboration; and to Grace Zisk, the technical editor of the paper and the conference proceedings.

References

ASIS (2003). *Emergency Planning Handbook,* 2nd Edition. Disaster Management Council, American Society for Industrial Security, Arlington, VA.

ASIS International (2003). *Counterterrorism and Contingency Planning Guide,* 2nd Edition. American Soc. for Industrial Security International, Arlington, VA.

Amin, M. (2002). "Toward secure and resilient interdependent infrastructures." Editorial, *Journal of Infrastructure Systems,* 8(3), 67-75.

Ardekani, S.A. (1992). "Transportation operations following the 1989 Loma Prieta earthquake." Eno Foundation for Transportation, Inc., *Transportation Quarterly*, 46(2), 219-233.

Arquilla, J. and D. Ronfeldt (Eds.) (2001). *Networks and Netwars: The Future of Terror, Crime, and Militancy*. Rand Report, Rand Corporation, Santa Monica, CA.

Barabási, A.L. (2002). *Linked: The New Science of Networks*. Cambridge, MA: Perseus Publishing.

Buckley, F. and M. Lewinter (2002). *A Friendly Introduction to Graph Theory*. New Jersey: Pearson Education, Inc.

Erickson, P.A. (1999). *Emergency Response Planning for Corporate and Municipal Managers*. American Society for Industrial Security, Arlington, VA.

Ezell, B. and Farr, J. (2000). "Infrastructure risk analysis model." *Journal of Infrastructure Systems*, 6(3), 114-117.

Gigliotti, R. and R. Jason (1991). *Emergency Planning for Maximum Protection*. American Society for Industrial Security (ASIS), Arlington, VA.

Haimes, Y.Y. (1998). *Risk Modeling, Assessment, and Management*. New York: John Wiley and Sons, Inc.

Haimes, Y.Y. and P. Jiang (2001). "Leontief-based model of risk in complex interconnected infrastructures." *Journal of Infrastructure Systems*, 7(1), 1-12.

Haimes, Y.Y. (1981). "Hierarchical holographic modeling." *IEEE Transactions on Systems, Man, and Cybernetics*, 11(9), 606-617.

Haimes, Y.Y., S. Kaplan, and J.H. Lambert (2002). "Risk filtering, ranking, and management framework using hierarchical holographic modeling." *Risk Analysis*, 22(2), 383-397.

Hall, S. (1989). *Handbook of Systems Engineering*. New York: Pergamon Press.

Heaney, J., J. Peterka, and L. Wright (2000). "Research needs for engineering aspects of natural disasters." *Journal of Infrastructure Systems*, 6(1), 4-14.

Hecker, E., W. Irwin, D. Cottrell, and C.A. Weatherby (2000). "Strategies for improving response and recovery in the future." *Natural Hazards Review*, 1(3), 161-170.

Juhl, G. (1993). "FEMA develops prototype disaster planning and response system." Communications Channels, Inc., *American City and County*, 108(3).

Kaplan, S. and B.J. Garrick (1981). "On the quantitative definition of risk." *Risk Analysis*, 1(1), 10-27.

Kovel, J. (2000). "Modeling disaster response planning." *Journal of Urban Planning and Development*, 126(1), 26-38.

Lambert, J.H., Y.Y. Haimes, D. Li, R. Schooff, and V. Tulsiani (2001). "Identification, ranking, and management of risks in a major system acquisition." *Journal of Reliability Engineering and System Safety*, 72(3), 315-325.

Lambert, J. and C. Patterson (2002). "Prioritization of schedule dependencies in hurricane recovery of transportation agency." *Journal of Infrastructure Systems*, 8(3), 103-111.

Lambert, J.H. and P. Sarda (2003). "Risk analysis of terrorism scenarios by superposition of infrastructure networks." Submitted to the *Journal of Infrastructure Systems*.

Marburger, A. and B. Westfechtel (2002). "Graph-based reengineering of telecommunication systems." *Graph Transformation, First International Conference (ICGT) Proceedings*, 2505, 270-285.

Mondul, S. (1997). "ITS: transportation and communications in service to multi-agency emergency response." *Merging the Transportation and Communications Revolutions: Proceedings, 7th Annual ITS America Meeting and Exposition*, ITS America, Washington, DC.

Myers, K.N. (1999). *Manager's Guide to Contingency Planning for Disasters: Protecting Vital Facilities and Critical Operations*, 2nd Edition. American Society for Industrial Security, Arlington, VA.

Parentela, E.M. and S.S. Nambisan (2000). *Emergency Response (Disaster Management)*, S. Easa and Y. Chan, (Eds.). American Society of Civil Engineers, (ASCE), Reston, VA.

Sarda, P. (2003). *Risk Identification in Interdependent Networks*. Master of Science thesis, Department of Systems and Information Engineering, University of Virginia, Charlottesville, VA.

Sharit, J. (2000). "A modeling framework for exposing risks in complex systems." *Journal of Risk Analysis*, 20(4), 469-482.

Watts, D.J. (2003). *Six Degrees: The Science of a Connected Age.* New York: W.W. Norton and Co.

Wei, B. (1991). "A unified approach to failure mode, effects and criticality analysis." *Proceedings of the Annual Reliability and Maintainability Symposium,* IEEE Reliability Society, New York, NY.

Xu, J. (2001). *Topological Structure and Analysis of Interconnection Networks.* Boston: Kluwer Academic Publishers.

The Fuzzy Logic Paradigm of Risk Analysis

Istvan Bogardi[1] and Lucien Duckstein[2]

Abstract

The fuzzy logic formulation of risk analysis is presented with application to a flood risk management case. If uncertainty in any element of risk analysis (exposure, resistance, and consequences) is expressed as a fuzzy set, the corresponding risk will also be calculated as a fuzzy set. The probabilistic formulation of risk analysis has several difficulties, including: the management of the low-probability/high-consequence case, the selection of the probability models, the scarcity of statistically meaningful data, the uncertainty of the consequence functions, and the covariances involved.

Fuzzy logic formulation, as an alternative to statistical methods to define uncertainties, is applicable when all these difficulties must be dealt with simultaneously. A short review of the application of fuzzy logic illustrates the versatility of this approach. A simplified flood risk management example illustrates a case where probabilistic and fuzzy uncertainties are involved in exposure, resistance, and consequences; economic and ecological consequences are considered, and a tradeoff analysis can identify the preferable action.

Introduction

The purpose of the paper is to present the fuzzy logic formulation of risk analysis with application to a flood risk management case. It is common to distinguish four main elements of risk analysis [Haimes 1998]. These are: 1) Exposure, L, which may be represented as a natural hazard (e.g., wind load) or human threat (e.g., terrorist attack), 2) Capacity/Resistance, R (e.g., strength of a bridge), 3) Failure event, $L > R$, and 4) Consequences of the failure (e.g., economic, human life, ecological, etc.). Risk

---

[1]Professor, Department of Civil Engineering, University of Nebraska-Lincoln, W359 Nebraska Hall, Lincoln, NE 68588-0531; 402-472-1726; ibogardi@unl.edu.
[2]Professor, Ecole Nationale du Genie Rural des Eaux et des Forets, 19 Avenue du Maine, 75732 Paris CEDEX 15, France; phone: (+33-0)-1-4549-8931; duckstein@engref.fr.

analysis is necessary in cases where uncertainties are inherent in any of the four elements. The traditional probabilistic formulation considers these uncertainties as random variables, represented, for instance, by the respective pdf: $g(L)$ and $f(R)$. Then the probability of failure, P $(L > R)$, can be calculated from $g(L)$ and $f(R)$. The economic consequences of failure can be represented by the loss function $D(L,R)$, and human health consequences by the dose-response relationship $DR(L,R)$, where DR is the probability of an individual developing the actual consequence, say cancer, given the exposure dose L and a possible nonzero threshold dose R.

Using the probabilistic formulation, the risk is commonly expressed as an expected value. Specifically, the engineering risk ER corresponds to the expected annual economic losses:

$$ER = \int_{R}^{\infty} D(L)\, g(L)\, dL$$

The human health risk then can be expressed as an expected probability:

$$HR = \int_{R}^{\infty} DR(L)\, g(L)\, d(L)$$

This probabilistic formulation has several known shortcomings, including:
(1) the management of the low-failure-probability/high-consequence case may be misrepresented by the expected value;
(2) the selection of the probability model, here the two pdf, is often arbitrary, while the results may be quite sensitive to this choice;
(3) statistically meaningful data on exposure and/or resistance are often lacking;
(4) the consequence functions, for instance flood losses in the domain of extreme floods or the dose-response function in the low-dose domain, are commonly quite uncertain.
(5) The covariances among the various types of exposures and resistances and the parameters involved in their estimation are commonly unknown. Again, the results are highly sensitive in this respect.

There are various methods to deal with some of these shortcomings. The conditional expected value formulation [Haimes 1998] helps to define the low-probability/high-risk dilemma in a more appropriate way. The probability bounds analysis [Ferson et al. 2002] accounts for the uncertainty of the probability model selection and the unknown covariances. A Bayes formulation may be used if data are unavailable for frequency interpretation [e.g., Carlin and Louis 2000]. The fuzzy set formulation can be considered as a practical alternative for these methods, and it is applicable when all of the above shortcomings must be dealt with simultaneously.

Fuzzy Sets and Water Resources Applications

Fuzzy logic is an alternative to statistical methods for defining uncertainties. In that respect, fuzziness represents situations where membership in sets cannot be defined on a yes/no basis because the boundaries of the sets are vague. The central concept of fuzzy-set theory is the membership function, which represents numerically the degree to which an element belongs to a set. As the degree increases to which an element belongs to a set, the value of the membership function for the element also increases.

In a classical set, a sharp or unambiguous distinction exists between the members and non-members of the set. In other words, the value of the membership function of each element in the classical set is either *1* for members (those that certainly belong to the set) or *0* for nonmembers (those that certainly do not). However, it is sometimes difficult to make a sharp or precise distinction between the members and nonmembers of a set. For example, the boundaries of the sets of highly contaminated water, deep wells, or numbers much greater than 1.0 are fuzzy.

Since the transition from member to nonmember appears gradual rather than abrupt, the fuzzy set introduces vagueness (with the aim of reducing complexity) by eliminating the sharp boundary that divides members of the set from nonmembers [Klir and Yuan 1995]. Thus, if an element is a member of a fuzzy set to some degree, the value of the function can be bounded, say between 0 and 1. When the membership function of an element only can have values 0 or 1, the fuzzy-set theory reverts to the classical-set theory. The membership value of a real number reflects the "likeliness" of the occurrence of that number; the level sets (intervals in this case) reflect different sets of numbers with a given minimum likeliness [Zimmermann 1991]. Any real number can be regarded as a fuzzy number and often is called a *crisp* number in fuzzy mathematics. The simplest type of fuzzy number is triangular, that is, linear on either side of the peak. Figure 1 gives an example of a triangular fuzzy number (TFN).

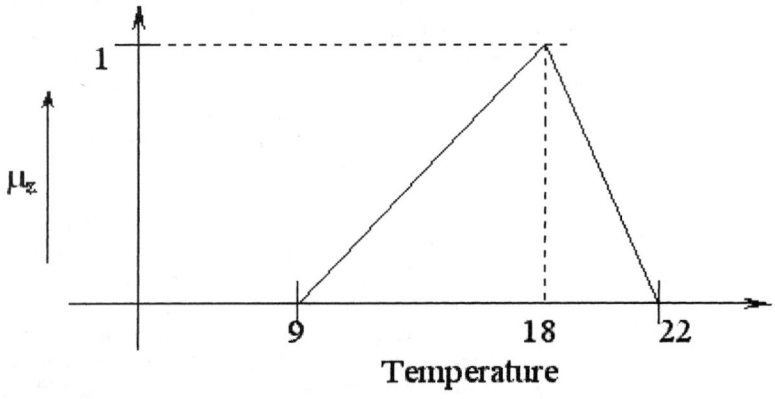

Figure 1. Membership function of temperature "higher than 9, lower than 22, around 18."

The fuzzy number displayed in Figure 1 can be described by the values of x at the points $a_1 = 9$, $a_2 = 18$ and $a_3 = 22$. Thus, $A = (a_1, a_2, a_3)$ completely characterizes the number.

Since the early application of fuzzy logic to water resources [Bogardi et al. 1983], there has been a great deal of research, and at present fuzzy logic has become a practical tool in water resources analysis and decisionmaking. Following are the main areas of application:

1. Fuzzy regression: this is useful when it is known that a causal relation exists but only very few data points are available [Bardossy et al. 1990, 1991b; Ozelkan and Duckstein 2000];

2. Hydrologic forecasting, for instance to embed short-term flood forecasting into medium-term forecasting. Kalman filtering is used for the short-term component, while fuzzy logic operates on the medium term, leading to a complete real-time forecasting system [Kojiri 1988];

3. Hydrologic modeling, where traditional rainfall runoff models can be replaced by fuzzy-rule systems with similar performance [Hundecha et al. 2001].

4. Fuzzy-set geostatistics allows us to use imprecise and possibly indirect measurements and small data sets in spatial statistical analysis [Bardossy et al. 1988, 1990];

5. Incorporation of spatial variability into groundwater flow and *transport modeling* with fuzzy logic [Dou et al. 1995, 1997a; Woldt et al. 1997]. In this approach, the imprecision of hydraulic parameters is embedded directly into the governing differential equations as fuzzy numbers. Then the system of finite difference equations is solved using fuzzy-set theory methods. This fuzzy modeling technique can handle imprecise parameters in a direct way without generating a large number of realizations (which is the common stochastic approach).

6. Regional water resources management aims at selecting among alternative management schemes under small data sets and imprecisely known or modeled objectives [Nachtnebel et al. 1986; Bardossy et al. 1989].

7. Multicriterion decisionmaking (MCDM) under uncertainty is essential 1) when water resources systems face multiple and conflicting criteria (objectives), e.g., economic efficiency and environmental preservation, and 2) when the criteria corresponding to alternative systems are imprecisely known [Duckstein et al. 1988; Bardossy et al. 1992; Bogardi et al. 1996]. These criteria are defined as fuzzy numbers and MCDM is performed in a fuzzy logic framework.

8. Fuzzy rule-based modeling has been used in several areas of hydrology, including the classification of spatial hydrometeorological events [Bardossy et al. 1995], climatic modeling of flooding [Bogardi et al. 1995], modeling of groundwater flow and transport [Bardossy and Disse 1993; Dou et al. 1997b, 1999; Woldt et al. 1997], regional scale nitrate leaching [Bardossy et al. 2002; Haberlandt et al. 2002], forecasting pollutants transport in surface waters [DiNatale et al. 2000], hydro-climatic modeling of hydrological extremes, i.e., droughts and intensive precipitation [Pesti et al. 1996; Pongracz et al, 2001].

9. Reservoir operation planning may apply fuzzy logic to derive operation rules [Simonovic 1992; Shrestha et al. 1996]. Operation rules are generated on the basis of economic development criteria such as hydropower; municipal, industrial, and irrigation demands; flood control and navigation; and environmental criteria such as water quality for fish and wildlife preservation, recreational needs, and downstream flow regulation. Split sampling of historical data (mean daily time series of flow, lake level, demands, and releases) is used to train and then validate the fuzzy-logic model. Such models appear to be easy to construct, apply, and extend to a complex system of reservoirs [Teegavarapu and Simonovic 1999].

10. Fuzzy risk analysis considers uncertainty in any or all elements of risk analysis: exposure or load, resistance or capacity, and consequence [Bogardi et al. 1989; Duckstein and Bogardi 1991]. The uncertainties are defined as fuzzy numbers, so the risk is also obtained as a fuzzy number. In a risk management framework, management options are evaluated to identify the best option, say in a risk-cost tradeoff formulation [Lee et al. 1994, 1995; Stansbury et al. 1999; Mujumdar and Sasikumar 2002]. This last application area is discussed in the following sections.

Fuzzy Logic Formulation of Risk

If uncertainty in any element of risk analysis is expressed as a fuzzy set, the corresponding risk also will be calculated as a fuzzy set. Among the numerous possibilities are these two examples:

Fuzzy engineering risk: If the exposure L is probabilistic, the resistance R is a fuzzy number, and the consequence function D is also fuzzy, then the engineering risk ER will be obtained as a fuzzy number:

$$E\widehat{R} = \int_R^\infty \widehat{D}(L)g(L)d(L)$$

Fuzzy health risk: If the exposure L is fuzzy, the resistance (threshold) $R=0$, and the dose-response relationship DR is fuzzy, then the health risk HR will be a fuzzy number, here the fuzzy probability:

$$H\widehat{R} = D\widehat{R}(\widehat{L})$$

This formulation was used by Bardossy et al. [1991a] to calculate the health risk of nitrate exposure. In this case, due to the immense uncertainty and lack of data, the dose-response relationship of nitrate exposure was characterized by fuzzy regression. That is, for each possible exposure, the probability of a health response, here cancer, appears as a fuzzy number. The numerical calculation of the fuzzy risk (i.e., to calculate a function of fuzzy numbers) is a relatively simple task that can be performed by several tools, such as the vertex method, for example [Dong and Shah 1987].

Often, a mixed probabilistic/fuzzy-set formulation is the most appropriate way to encode uncertainties. The next section provides such an example; it also indicates that the fuzzy-risk formulation is amenable to risk management purposes.

Flood Management Example

A reach of the Santa Cruz River in southern Arizona is used to illustrate an application of the mixed probabilistic/fuzzy-set risk management approach. This simplified case consists of a binary decision problem: to protect or not to protect against the maximum peak floods.

The four main elements of this flood risk analysis include:
1. Flood hazard as the exposure—here the flood discharge exceedance L,
2. Resistance, or capacity of flood-control works R,
3. Failure event, whenever $L > R$, and
4. Consequences of the binary decision—here, the economic losses of flooding, the cost of protection, and the ecological impacts of human intervention (protection) are considered.

The problem is formulated using:
  (a) Two states of nature: flooding S_1, when $L \geq R$ and S_2: $L < R$,
  (b) A Bernoulli model of occurrence of state S_i, i = 1,2:
  $$P(S_1) = P(L \geq R) = p$$
  $$P(S_2) = P(L < R) = 1 - p$$
  (c) Two actions: $a_1 = $ *protect* and $a_2 = $ *do not protect*, and
  (d) Annualized flood losses K and protection costs C.

The following uncertainties involved in the elements of the risk analysis are represented by triangular fuzzy numbers:
1. Exceedence probability: $p = (7, 8, 10) \times 10^{-4}$
2. Annual economic losses: $K = (4, 5, 6) \times 10^6$
3. Annual protection costs: $C = (3, 4, 5) \times 10^3$, a cost of "about" 4000

The economic consequences of the two actions are shown in the following economic loss matrix:

I	S_i	a_1 protect	a_2 do not	$p(S_1)$
1	$L > R$	C	K	p
2	$L \leq R$	C	0	1-p

The expected annual economic losses D, corresponding to actions a_1 and a_2 are:

$$D(a_1) = C(.) p (+) C(.)(1-p) = C$$
$$D(a_2) = K(.) p$$

Fuzzy arithmetic is used to calculate $D(a_1)$ and $D(a_2)$ as fuzzy numbers to obtain:

$$D(a_1) = (3,4,5) \times 10^3 \text{ and } D(a_2) = (2.8, 4, 4.6) \times 10^3$$

To compare $D(a_1)$ and $D(a_2)$, the fuzzy mean, that is, the center of gravity of the fuzzy numbers is used:

$$D(a_1) = 4 < D(a_2) = 4.27.$$

Thus, on a purely economic basis, action $a = a_1$, *protect*, is preferred to action a_2

The ecological effect of flood protection is measured by the number of species lost if a flood-control structure is built. The initial number of the species is assumed to be known precisely: $N_1 = 60$. Diversity after construction, which is not predictable precisely, is represented by a fuzzy number: $N_2 = N(a_2) = (10, 25, 40)$, corresponding to about 25 species left, but not less than 10 and not more than 40.

The ecological consequences of the two actions are shown in the following ecological matrix:

i	S_i	a_1 protect	a_2 do not protect	$p(S_i)$
1	$L > R$	N_2	N_1	p
2	$L \leq R$	N_2	N_1	$1-p$

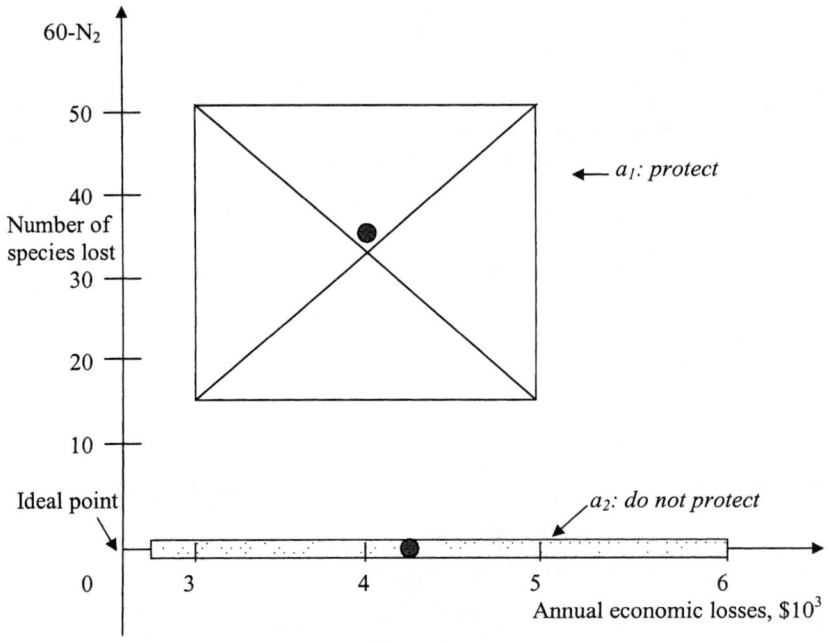

Figure 2. Fuzzy tradeoff analysis.

To evaluate the two actions a_1 and a_2 from a joint economic/ecological aspect, a fuzzy tradeoff relationship is constructed (Figure 2). Action a_1 *(protect)* is represented by a two-dimensional fuzzy number reflecting an uncertain economic and ecological quantity. Action a_2 *(do not protect)* does not show ecological uncertainty and is thus represented by a one-dimensional fuzzy number projected as the horizontal segment *(2.8,4,6)*. The action closest to the ideal point—neither economic nor ecological losses—is selected: the distance between two fuzzy numbers (here, one point and one fuzzy number) thus must be defined. Taking again the mean value representation for a fuzzy number, it is found that action a_2 is preferred to action a_1. Thus, taking ecological effect into account leads to a reversal of the decision that would be taken on purely economic grounds.

Conclusions

1. One of the elements of risk analysis, exposure, may include natural hazards and human-caused threats.
2. The probabilistic formulation of risk analysis may have difficulties when statistical data are unavailable. Examples are human-caused threats, cases when several consequences are considered, and when the elements are involved (e.g., exposures and resistances are statistically dependent).
3. Fuzzy-logic formulation is an alternative to the probabilistic formulation.
4. A simplified flood-risk management illustrates a case where:
    ⇒ probabilistic and fuzzy uncertainties are involved in exposure, resistance, and consequences,
    ⇒ economic and ecological consequences are considered, and
    ⇒ a tradeoff analysis can identify the preferable action.

References

Bardossy, A. and M. Disse (1993). "Fuzzy rule-based models for infiltration." *Water Resources Research*, 29, 373-382.

Bardossy, A., I. Bogardi, and W.E. Kelly (1988). "Imprecise (fuzzy) information in geostatistics." *Mathematical Geology*, 1(4), 287-311.

Bardossy, A., I. Bogardi, L. Duckstein, and P. Nachtnebel (1989). "Fuzzy decision-making to resolve regional conflicts between industry and the environment." In Evans, C.W., W. Karwowski, and P.M. Wilhelm, (Eds.), *Fuzzy Methodologies for Industrial and Systems Engineering*, Chapter 3. Amsterdam: Elsevier.

Bardossy, A., I. Bogardi, and L. Duckstein (1990). "Fuzzy regression in hydrology." *Water Resources Research*, 25(7), 1497-1508.

Bardossy, A., I. Bogardi, and L. Duckstein (1991a). "Fuzzy set and probabilistic techniques for health-risk analysis." *Applied Mathematics and Computation*, 45(3), 241-268.

Bardossy, A., R. Hagaman, L. Duckstein, and I. Bogardi (1991b). "Fuzzy least squares regression: theory and application." In Fedrizzi, M. and J. Kacprzyk, (Eds.), *Fuzzy Regression Models*, 66-86. Warsaw, Poland: Omnitech Press.

Bardossy, A., L. Duckstein, and I. Bogardi (1992). "Fuzzy composite programming with water resources engineering application." In: *Proc. Fourth World Congress of the International Fuzzy System Association*, Brussels, Belgium.

Bardossy, A., L. Duckstein, and I. Bogardi (1995). "Fuzzy rule-based classification of circulation patterns for precipitation events." *Intern. J. of Climatology*, 15(10), 1087-1097.

Bardossy, A., U. Haberlandt, and V. Krysanova (2002). "Automatic fuzzy-rule assessment and its application to the modeling of nitrogen leaching for large regions." *Soft Computing*, accepted for publication.

Bogardi, I., L. Duckstein, and A. Bardossy (1983). "Regional management of an aquifer under fuzzy environmental objectives." *Water Resources Research*, 19(6), 1394-1402.

Bogardi, I., L. Duckstein, and A. Bardossy (1989). "Uncertainties in environmental risk analysis." In Haimes, Y.Y. and E.Z. Stakhiv, (Eds.), *Risk Analysis and Management of Natural and Man-made Hazards*, 342-356. New York: ASCE.

Bogardi, I., R. Reitel, and P. Nachtnebel (1995). "Fuzzy rule-based estimation of flood probabilities under climatic fluctuation." In Haimes, Y.Y., D.A. Moser, D.A., and E.Z. Stakhiv, (Eds.), *Risk-Based Decision Making in Water Resources VII*, 61-79. Reston, VA: ASCE.

Bogardi, I., A. Bardossy, and L. Duckstein (1996). "Conflict analysis using multiple criterion decision making under uncertainty." In Ganoulis, J. (Ed.), *Transboundary Water Resources Management: Theory and Practices*, 79-98. Heidelberg: Springer-Verlag.

Carlin, B.P. and T.A. Louis (2000). *Bayes and Empirical Bayes Methods for Data Analysis*. Chapman and Hall, 419.

DiNatale, M., L. Duckstein, and A. Pasanisi (2000). "Forecasting pollutants transport in river by a fuzzy rule-based model." In *Workshop on Fuzzy Logic and Applications*. Belgium: Mons Institute of Technology.

Dong, W. and H.C. Shah (1987). "Vertex method for computing functions of fuzzy variables." *Fuzzy Sets and Systems*, 24, 65-78.

Dou, C., W. Woldt, I. Bogardi, and M. Dahab (1995). "Steady state groundwater flow simulation with imprecise parameters." *Water Resources Res.*, 31(11), 2709-2719.

Dou, C., W. Woldt, M. Dahab, and I. Bogardi (1997a). "Transient groundwater flow simulation using a fuzzy set approach." *Ground Water,* 35(2), 205-215.

Dou, C., W. Woldt, I. Bogardi, and M. Dahab (1997b). "Numerical solute transport simulation using fuzzy sets approach." *Journal of Contaminant Hydrology,* 27(1-2), 107-126.

Dou, C., W. Woldt, and I. Bogardi (1999). "Fuzzy rule-based approach to describe solute transport in the unsaturated zone." *J. of Hydrology,* 220(1-2), 74-85.

Duckstein, L. and I Bogardi (1991). "Reliability with fuzzy elements in water quantity and quality problems." In Ganoulis, J. (Ed.), *Risk and Reliability in Water Resources and Environmental Engineering,* 78-99. Berlin: Springer Verlag.

Duckstein, L., P. Korhonen, and A. Tecle (1988). "Multiobjective forest management using a visual, interactive and fuzzy approach." In *Proc., 1988 Symposium on Systems Analysis in Forest Resources,* 68-74. USDA Forest Service, Fort Collins, CO.

Ferson, S., L. Ginzburg, J. Hajagos, and W.T. Tucker (2002). "Terrorism's implications for uncertainty calculi." *Risk-Based Decisionmaking in Water Resources X,* Engineering Foundation Conference, Santa Barbara, CA.

Haberlandt, U., V. Krysanova, and A. Bardossy (2002). "Assessment of nitrogen leaching from arable land in large river basins, Part II: regionalization using fuzzy rule-based modeling." *Ecological Modeling,* 150(3), 277-294.

Haimes, Y.Y. (1998). *Risk Modeling, Assessment, and Management,* New York: John Wiley and Sons, 726.

Hundecha, Y., A. Bardossy, and H.W. Theisen (2001). "Development of a fuzzy logic-based rainfall-runoff model." *Hydrological Sciences Journal,* 46(3), 363-376.

Klir, G. J. and K. Yuan (1995). *Fuzzy Sets and Fuzzy Logic: Theory and Application.* Englewood Cliffs, NJ: Prentice Hall.

Kojiri, T. (1988). "Real-time reservoir operation with inflow prediction by using fuzzy inference theory." In *Seminar on Conflict Analysis in Reservoir Management,* Session F. Asian Institute of Technology, Bangkok, Thailand, December 1988.

Lee, Y.W., M.F. Dahab, and I. Bogardi (1994). "Fuzzy decision making in ground water nitrate risk management." *Water Resources Bulletin,* 30(1), 135-148.

Lee, Y.W., M.F. Dahab, and I. Bogardi (1995). "Nitrate risk assessment using a fuzzy-set approach." *Journal of Environmental Engineering,* 121(3), 245-256.

Mujumdar, P.P. and K. Sasikumar (2002). "A fuzzy risk approach for seasonal water quality management of a river system." *Water Resources Research*, 38(1), 55-63.

Nachtnebel, H.P., P. Hanish, and L. Duckstein (1986). "Multicriterion analysis of small hydropower plants under fuzzy objectives." *The Annals of Regional Science*, XX, 86-100.

Ozelkan, E.C. and L. Duckstein (2000). "Multi-objective fuzzy regression: a general framework." *Computers and Operations Research*, 27, 7-8, 635-652.

Pesti, G., B. Shrestha, L. Duckstein, and I. Bogardi (1996). "A fuzzy rule-based approach to drought assessment." *Water Resources Research*, 32(6), 1741-1747.

Pongracz, R., J. Bartholy, and I. Bogardi (2001). "Fuzzy rule-based prediction of monthly precipitation." *Physics and Chemistry of the Earth, Part B.*, 26(9), 663-667.

Shrestha, B.P., L. Duckstein, and E.Z. Stakhiv (1996). "Fuzzy rule-based modeling of reservoir operation." *Journal of Water Resources Planning and Management*, 122(4), 262-269.

Simonovic, S.P. (1992). "Reservoir systems analysis—closing the gap between theory and practice." *Journal of Water Resources Planning and Management*, 118(3), 262-280.

Stansbury, J., I. Bogardi, and E.Z. Stakhiv (1999). "Risk-cost optimization under uncertainty for dredged material disposal." *Journal of Water Resources Planning and Management*, 125(6), 342-351.

Teegavarapu, R.S.V. and S.P. Simonovic (1999). "Modeling uncertainty in reservoir loss functions using fuzzy sets." *Water Resources Research*, 35(9), 2815-2823.

Woldt, W.E., C. Dou, and I. Bogardi (1997). "Innovations in modeling solute transport in the vadose zone using fuzzy rule-based methods." In: *Proceedings of ASAE Conference: Emerging Technologies in Hydrology*, 2097-2098. ASAE.

Zimmermann, H.J. (1991). *Fuzzy Set Theory and Its Applications*. Boston, MA: Kluwer-Nijhoff.

A Dynamic Risk Model for Information Technology Security in a Critical Infrastructure Environment

John H. Saunders[1]

Abstract

The risk assessment, modeling, and simulation of *critical infrastructure* information technology (IT) security has been limited to broad, macro-level approaches. Concurrently, risk assessment in IT security has been limited to static analysis and modeling. This paper provides a dynamic risk framework and a model that synthesizes elements of an organizational decision model on both macro and micro levels. In the proposed dynamic model the focus is upon building the reactive capability of an infrastructure organization as well as preparing for time-based cascading effects.

Introduction

This paper begins by describing threats to the infrastructure community in information technology security. It then relates countermeasures for reducing those threats. A framework is proposed for balancing the countermeasures against the threats over time, given limited resources. Finally, a practical model is demonstrated which allows the architect of an information security risk policy to target a desired level of risk.

Information Security Threats

How difficult is it for a person or persons with evil intent to break into the IT system of an infrastructure provider such as an electrical utility, a transportation control center, or a health provider? We know that no IT system is infallible. There are

---

[1] Professor, National Defense University, Washington, DC 20319; (202) 685-2078; saunders@ndu.edu.

hundreds, if not thousands, of known methods[2] and tools for attacking information technology systems. These methods exploit vulnerabilities, i.e., weaknesses in hardware, software, and people. Examples of vulnerabilities in these areas include communications line taps, software buffer overflows, remote procedure calls, poor passwords, and untrained system administrators. Many tools and checklists are available for discovering these vulnerabilities. Unfortunately, the same tools available to the "white hats" are also readily procurable by the "black hats."

Examples of these tools include scanners, script kiddy tools, sniffers, and rootkits. A scanner tool such as *SuperScan* looks for open "ports" (like open windows in a house) on a machine such as a network server. When a port is found the attacker is alerted and can then use the open port to explore the machine. If there is a program on the machine such as the tiny file transfer program (tftp.exe—often installed by default with the computer), the intruder can deposit a "Trojan horse" piece of software such as *Back Orifice 2K (BO2K)*. Pieces of software such as *BO2K* are known as "script kiddy" tools because they are simple to use. Even a fairly naïve user can craft harmful attacks with them. An insider can run a sniffer such as *Ethereal* from any computer on a network. The program is configurable to sit and wait for a user to log on; then it can grab the password[3]. A root kit such as *Linux Rootkit 5* is used on computers to replace common low-level routines with bogus copies that have been modified to report information back to their propagator or report false information to the valid user. The more sophisticated rootkits can redirect calls to other programs and then cover their tracks. The notorious "malware" incidents such as the ILOVEU, Code Red, and NIMDA "worm"-type viruses exploited vulnerabilities in e-mail systems. There are toolkits for constructing these types of viruses as well.

Most of these vulnerability discovery tools have valid uses for network engineers. The same tools used by the hacker may be used to help engineers measure traffic, pull out certain kinds of digital traffic for more efficient routing, or discover security leaks in networks. Other tools such as virus creation toolkits are marketed as aides to help security engineers learn how vulnerabilities are exploited. Complicating the job of the network security engineer is the daily release of new software and hardware. New releases yield new targets. At the same time, many systems give out "telling" information about their configuration by default. Prior to early 2002, versions of the Microsoft Windows 2000 Server automatically opened up many very

---

[2] Newer attacks are assembled by gathering information from a variety of sources and capitalizing upon weaknesses in a number of areas. Much of this process has been laid out in the book *Hacking Exposed* [McClure et al. 2001]. This text outlines the steps of footprinting, scanning, enumeration, gaining access, escalating, privilege, pilfering, covering tracks, creating back doors, and finally, launching attacks.

[3] An important concept to understand is that the original design of wired (Ethernet) and wireless (e.g., IEEE 802.11 standard) computer networks capitalizes upon the broadcasting of all messages to all users. This is the default method. Therefore, any user connected in any way to that network segment can "hear" all that is transpiring.

vulnerable services, such as the *Internet Information Service* and the *Domain Name Service*. Installing this software was akin to living in a home without any locks on the doors or windows.

For more detail about these specific methods and tools, refer to the SANS Institutes Reading Room at http://rr.sans.org. Or you may also visit *Computer Security Links* at http://www.johnsaunders.com/security.htm. There you will find links to categories of attacks as well as links to databases filled with vulnerabilities. Another excellent resource for learning about vulnerability analysis is the National Institute for Science and Technology's publication 800-42 *Guideline on Network Security Testing* [NIST 2002].

The Challenge: Specific Vulnerabilities and Threats to Infrastructure Sensor and Control Systems

Given that many threats to general networks exist, what impact might this have on specific types of control systems (CS) such as supervisory control and data acquisition (SCADA) networks used by critical infrastructure organizations? Aren't these systems typically more isolated, with specific vs. general purposes? In this paper, henceforth we will refer to entire "operational" systems used by infrastructure organizations as *SCADA*. In truth, there are many IT elements under this umbrella. These include sensors, valves, logic controllers, switches, intermediate "intelligent" devices, and communications lines, as well as full-scale computers and computer networks. Since the early 1990s there have been many movements toward standardizing protocols[4] for SCADA systems. Examples include the Utility Communication Architecture (UCA), the International Electro-technical Committee's (IEC) selection of TCP/IP as the mandatory networking protocol for intra- and inter-substation communications, and use of the Manufacturing Messaging Specification (MMS) as an application layer standard for services to read, write, define, and create data objects. It is important to understand that this standardization movement presents a new set of risks to the critical infrastructure community.

There are two general areas where different types of vulnerabilities exist in control systems. The first is in legacy systems, which are patchworks of controls put in place over decades of IT materialization. The second is the newer emergent systems that are focused upon standard configurations. The good news for the much older legacy control equipment is that these systems are relatively safe from attack by outsiders. The logic in these systems tends to be hardwired and/or proprietary. However these systems do tend to be more vulnerable in terms of physical exposure; they are located at remote sites in boxes that may be easily compromised or accessed through easily accessible low-bandwidth modems. The greater threat may be from insiders who understand the simpler hardwired nature of the equipment and have ready access.

---

[4] A protocol is a standard for the way computer and communications hardware, software, and data interact.

As in any industry, IT has provided the infrastructure industries with emergent options for doing business. Table 1 below provides an insight into the emerging IT-related technology along with the associated vulnerability and threat agent.

Table 1. Computer security challenges in the SCADA environment.

Occurring in infrastructure industries	Occurring in information security (IS)
Quantity and extent of supervisory control and data acquisition (SCADA) systems	Most-hacked: critical infrastructures
Move toward standardized platforms, e.g., Windows 2000	Most vulnerabilities: Microsoft, esp. web-based software
Move toward internet protocol (IP) and simple network mgmt. protocol (SNMP) as a base for communication and management	Easiest protocols to hack: IP and SNMP
Move toward greater connectivity, especially to business-type networks, e.g., for billing	Opportunities for access: increasing
Wireless connectivity	Best mode for easy and unprotected access: wireless
Constrained and highly controlled IT resources	Information security resources: very complex, difficult to manage, significant emerging technology
Difficulty in locating IT personnel	Of all IT professionals: security professionals rare, expensive

Many of the IT changes occurring in the infrastructure arena are good business decisions and provide benefits to the user community. Using IT to help sense and/or control the flow of water or electricity provides efficiency and economy in operations. A move toward common IT platforms such as Microsoft Windows running on Intel chips allows an organization to quickly install computer programs which integrate components of their infrastructure. At the same time, using common data communications protocols such as the internet protocol (IP) and the simple network management protocol (SNMP) also allows infrastructure components to communicate freely.

However, as these changes occur, they introduce new vulnerabilities. Some of the vulnerabilities have been highlighted by Dr. Samuel Varnado of Sandia Labs[5]:

---

[5]Varnado, Dr. Samuel G., Director of Sandia National Laboratories' Infrastructure and Information Systems Center. Statement to United States House of Representatives, Committee on Energy and Commerce, Subcommittee on Oversight and Investigations, July 9, 2002.

"Sandia has been investigating vulnerabilities in SCADA systems for five years. During this time, many have been found. Our assessments show that security implementations are, in many cases, non-existent or based on false premises. Some of the vulnerabilities in legacy SCADA systems include inadequate password policies and security administration, no data protection mechanisms, and information links that are prone to snooping, interruption, and interception. When firewalls are used, they are sometimes not adequately configured, and there is often "backdoor" access because of connections to contractors and maintenance staff. We have found many cases in which there is unprotected remote access that circumvents the firewall. From a security perspective, it should be noted that most of the SCADA manufacturers are foreign-owned. In summary, it is possible to covertly and easily take over control of one of these systems and cause disruptions with significant consequences. Recognition of that fact led numerous federal agencies and municipal water and transportation systems to request Sandia help following September 11. Of even more concern is the fact that the control systems are now evolving to the use of the internet as the control backbone. The electric power grid is now, under restructuring, being operated in a way for which it was never designed. More access to control systems is being granted to more users; there is more demand for real-time control; and business and control systems are being connected. Typically, these new systems are not designed with security in mind. More vulnerabilities are being found, and consequences of disruptions are increasing rapidly. Industry is now asking for our help in understanding vulnerabilities, consequences, and mitigation strategies."

Specific lists of vulnerabilities for commercial off-the-shelf (COTS) equipment may be found at *CERT* or at *Security Focus*.[6] At these same locations you will find suggested methods for countering the highlighted vulnerability.

Available Countermeasures

While the bad news is that threats and threat agents are ubiquitous, the good news is that for every known vulnerability, there usually is, or soon will be, a matching countermeasure. The caveat is that not all vulnerabilities are known. Table 2 below outlines 31 general classes of countermeasures that may be used to fight specific vulnerabilities or as a combined general fortress against possible new attacks.

A number of general recommendations for combating SCADA vulnerabilities have been developed by the federal government [USDOE 2002]. Examples from these recommendations include: "Identify all connections to SCADA networks," "Disconnect unnecessary connections to the SCADA network," and "Establish strong controls over any medium that is used as a backdoor into the SCADA network." It seems apparent that more work needs to be done to identify those issues specifically associated with SCADA-type systems. As such, the federal government through the

---

[6] http://www.cert.com; http://www.securityfocus.com

National Institute of Standards and Technology (NIST) has commissioned the Process Control Security Requirements Forum (PCSRF)[7]. This group has members from government and industry supporting the development and dissemination of standards for process control security.

Table 2. Countermeasures for information security vulnerabilities.

People	Network Technology
• Formal written policy	• Firewalls/router security
• Background checks	• Intrusion-detection systems
• Incident response team	• Disconnect
• User safety and response training	• Integrity checking
Processes	• Honeypots
• Updating	**Encryption**
• Secure software configuration	• Digital certificates
• Backups	• Virtual private networks
• Log file analysis	• Database encryption
• Physical and environmental security	• Wireless equivalency protocol
	• Pretty good privacy (PGP) e-mail
Authentication and Access	**Management**
• Biometrics	• Adequate budget
• Passwords and tokens	• Effective personnel functions
• Database access control	• Contingency planning
• Server/segment access control	• System audit and vulnerability analysis
Computer Level	
• Antivirus protection	
• Web browser controls	
• Operating system controls	
• Redundant hardware or software	

The challenge facing an information security manager is to apply the best available set of countermeasures given available resources—people, technology, funding, and time. The manager must minimize *residual risk*, i.e., that portion of risk that remains after security measures have been applied. Unlike many other areas of IT management, an important consideration in managing information security is time.

The Importance of a Time Focus in Counteracting Threats

The importance of reaction time, automated or manual, tends to take on greater meaning in SCADA systems than in more traditional networks. Downtime and reaction time in an infrastructure organization may create critical, life-threatening situations. For some infrastructure components such as programmable logic controllers (PLC), a fraction of a second may be the unit of concern. That is the time required to toggle a switch. In an electrical system, a malfunctioning or misprogrammed switch may lead to the buildup of current and an eventual explosion. In many networked environments, the short-term time focus is the reason for such a

---

[7] http://www.isd.mel.nist.gov/projects/processcontrol/

large emphasis on automated countermeasures such as firewalls and intrusion-detection systems. These systems can be set up to check every incoming (and outgoing) piece of data on a "real-time" basis. If the signature of an attack is known, then the countermeasure can halt its progress and provide an alert to human operators.

There are time-frames beyond an emergency level that should be considered. For other components, such as placing a generator into service, the lead-time may be months or years. During that extended time an entire community may suffer. Table 3 provides a framework for better understanding the time-frame relationships.

Table 3. Time components in information security implementation.

Reference Time	Definition	Examples/Countermeasures
Emergency—seconds to hours	Time to deal with real-time emergencies—component and system crashes, denial-of-service attacks	Emergency response teams and procedures; backup systems; side-channel communications lines
Update—hours to days	Time to insure that announced vulnerability patches are routinely installed	Security patches to Windows 2000; insuring compatibility in device drivers when new equipment is installed
Long-term	Reacting so as to keep any potential unknown threats at a distance from the organization and its systems	Keeping critical information about systems away from public access; keeping operations centers in "quiet" locations; erecting multiple security hurdles—"defense in depth"

As threats are increasing, gaps in the three areas outlined above need to be closed. Given the importance of the time factor, it is essential that it be an integral part of understanding the relationship between threats, vulnerabilities, and countermeasures in information security modeling.

Risk Frameworks for Information Security

While it is clear that measures should be taken to reduce the risk imposed by the myriad threats, just "how" to accomplish this is more of a mystery. Regretfully, security is not large on the RADARscope for many organizations. As an overhead cost, it is often under-funded. Operators are not well-trained in security issues. Security crises are poorly handled because response teams have not been established.

The challenge for the IT and operations managers in this type of environment is to:
1. properly analyze the vulnerabilities and threats to an information system,
2. identify the potential impact on the business of a loss of information or system capabilities, and based upon these analyses,
3. identify appropriate and cost-effective countermeasures.

This challenge is known as *risk assessment*. Haimes [1998] and others have thoroughly explored the risk assessment process. The framework they have established allows practitioners to analyze and synthesize risk within the following macro-level framework:
- *Analysis*
  - What can go wrong?
  - What is the likelihood that it would go wrong?
  - What are the consequences?
- *Synthesis*
  - What can be done?
  - What options are available and what are their associated tradeoffs in terms of all costs, benefits, and risks?
  - What are the impacts of current management decisions on future options?

For assessing risk in infrastructure industries, models such as those proposed by Ezell et al. [2000] and by the Association of Metropolitan Sewerage Agencies (AMSA) [2002] fit well within this framework. A model that has been produced independently within the IT security community also works quite well [CCIMB 1999]. This semantic model is shown in Figure 1.

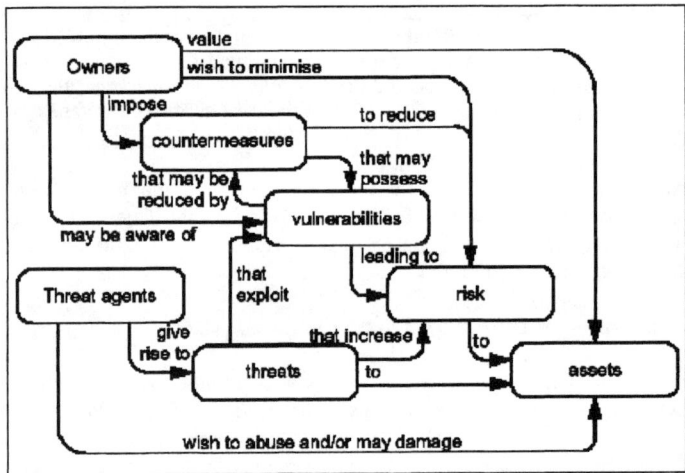

Figure 1. Common criteria security model.

This model was developed as part of the common criteria (CC) effort. The CC Board is an international standards body that has developed common standards for evaluating IT security products. It provides a good general framework, but the devil lies in the execution of its details.

Risk Assessment Challenges and Methods

Longstaff et al. [2000] aptly framed the risk assessment challenge with their statement, "To the extent risk assessment is precise, it is not real; to the extent risk assessment is real, it is not precise." In assessing the risk involved in information security, major questions such as the following need to be addressed:
1. What specific variables should be included in the model? How do we represent threats, vulnerabilities, assets, and countermeasures?
2. What level of detail is necessary?
3. What specific values do we attach to these variables?
4. Do these values change over time? If so, how?
5. How do we synthesize all the variables? Is there a single entity called *risk,* or multiple aspects? What algorithm is used to determine how we combine the representational elements of threat and countermeasure?

Ultimately, decisionmakers need assistance in determining which countermeasures have the greatest impact. If limited resources are available, how should they be applied to gain the greatest leverage? Do we upgrade our firewalls, do more software testing, apply more funds to security education, encrypt our databases, or do vulnerability testing? The list is endless. Further, which do we do first, and when?

There are a variety of detailed mathematical and/or heuristic approaches toward factoring these multiple objectives in assessing risk. A number of these methods are summarized for information security in the *International Critical Information Infrastructure Protection Handbook* [Wenger et al. 2002]; they are covered extensively for general application by Haimes [1998]. Some methods, which use *dollars* as an analytical base, include *loss expectancy, risk filtering and ranking with cost/benefit,* and *insurance-based actuarial* methods. Another general approach is to use a *relative metric* to measure the value of applying countermeasures to patch the vulnerabilities in systems. Using a relative metric has a drawback in that you must first have a base against which your metric can be compared. Specific methods for infrastructure organizations to evaluate their risk postures have been provided by Sandia Labs (RAM-D and RAM-T) for the Department of Energy [SANDIA 2002] and by the Association of Metropolitan Sewerage Agencies [AMSA 2002] for its membership. However, these accepted methods focus little on some of the more important aspects of information security modeling, including time, level of detail, and synthesis. Another method is needed to provide this added value.

A Selected Risk Model Example

A modeling methodology that has the capability to answer the issues of time, detail, and synthesis is called *System Dynamics (SD)* [Forrester 1961]. This method uses a flexible approach to represent an appropriate level of detail, to synthesize variables, and to incorporate the dynamics of changing model variables. Fundamentally, SD uses two connected entities: levels and rates, also known as stocks and flows. *Stocks* represent quantities of tangible and/or intangible entities. The symbol for a stock is a rectangle. Typical examples of stocks include people, dollars, computers, morale, attitude, and risk. *Flows* are equivalent to valves; that is, devices for setting how much quantity may flow into or out of a stock in a given time period.

The diagrams in Figure 2 below portray simple stock and flow scenarios. Stocks and flows run in a simulation environment where they can take on different values each period. For example, the level of people in a stock may be 100 in the first period and 110 in the second. The rate of change of flows into a stock may be a constant, (e.g., increasing at 10 units per period) or a variable (e.g., 20% of the previous period's level). Other constants or variables that may impact the environment are represented by circles.

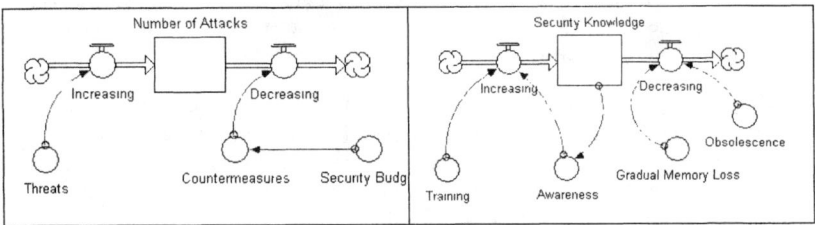

Figure 2. Stock and flow diagrams.

In the left diagram of Figure 2 above, the number of attacks will increase based upon a function of the number of threats. Conversely, the number of attacks expected each period will decrease based upon countermeasures put in place, such as firewalls, antivirus protection, and user training. The number of countermeasures put in place is affected by the organization's budget for security. In the right diagram, the security knowledge in an organization will increase as personnel are trained, and will decrease as users gradually forget what they have learned or as new software is installed and the previous knowledge becomes obsolete.

The values of the variable stocks and flows are computed each period through simultaneous difference equations. The equations for the *Number of Attacks* example follow in Figure 3.

```
Number_of_Attacks(t) = Number_of_Attacks(t - dt) + (Increasing - Decreasing) * dt
  INIT Number_of_Attacks = 100
  INFLOWS:
      Increasing = SIN(Threats)
  OUTFLOWS:
      Decreasing = Countermeasures
  Countermeasures = COS(Security_Budget)
  Security_Budget = 25
  Threats = GRAPH(TIME)
  (1.00, 20.0), (2.00, 30.0), (3.00, 40.0), (4.00, 45.0), (5.00, 50.0), (6.00, 55.0), (7.00, 60.0), (8.00, 55.0),
  (9.00, 50.0), (10.0, 45.0), (11.0, 40.0), (12.0, 35.0), (13.0, 30.0)
```

Figure 3. Relationship of variables.

These relationships would need to be defined by experts and by personnel with organizational knowledge. Typical instantiation would include, for example, that one-third of an end user's security knowledge declines over a one-year period. Also, that the use of a well-configured intrusion-detection system such as *Snort* contributes to an overall decrease in the number of attacks by 10% per year. While it could be argued that defining these relationships is difficult, this type of knowledge is necessary for an information security manager in order to justify his/her allocation of resources.

Output from an SD simulation is typically expressed by a *behavior over time* graph such as the one in Figure 4 below. This graph depicts the relationships among the variables of threats, attacks, and budget over a 12-month period. By viewing these relationships we can hypothesize what possible modifications may be beneficial to the system. In this rare case, it appears that the organization should consider reducing and varying the security budget to handle peak periods.

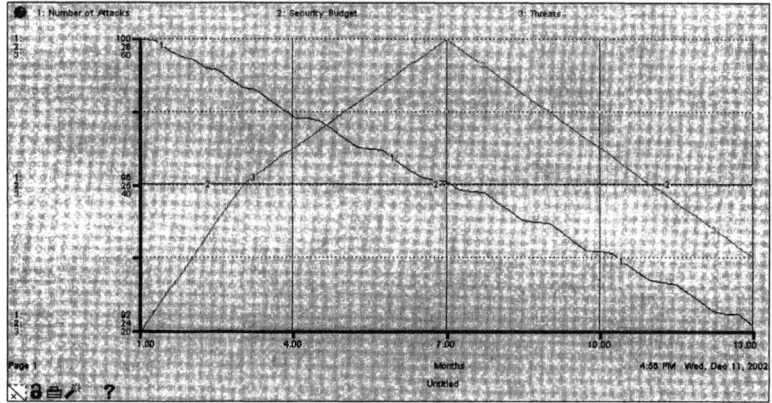

Figure 4. *Behavior over time* graph.

System Dynamics models may be expanded and then aggregated into "sectors" to allow a user and/or modeler to better understand the scope of the problem. The sectors in the macro-model below (Figure 5) mimic the common criteria (CC) semantic network displayed in Figure 1.

The time period of interest in an SD model, called *delta time (dt)*, is a flexible feature. This time period may be set as a micro-second or as a year, whatever best suits the particular scenario under observation. In a water system, for example, a delay of a few minutes in the opening or closing of a gate may have little overall effect upon the general system health, so the dt may be set as one minute. In an electric transmission system, such a delay may have serious consequences, including the buildup of current and an explosion; therefore, dt would need to be set at a much smaller value, such as one second.

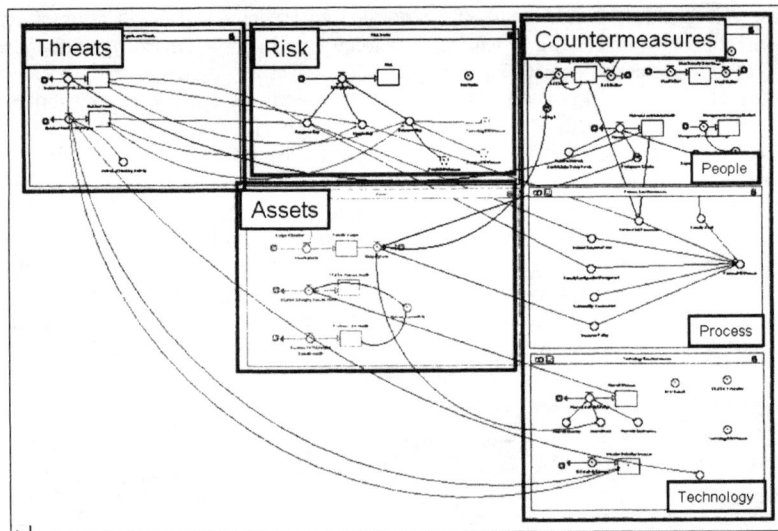

Figure 5. Infrastructure model.

Tangible representation. System Dynamics sectors can represent very specific, tangible entities and relationships. The example in Figure 6 below, an assets sector, provides a view of the interconnection between computer/SCADA network assets and physical valves and water levels. The upper portion portrays the electronic components of the system and the bottom portion portrays the physical assets. Feedback is shuttled among sensors, remote terminal units, and gates. The presence or absence of a firewall in this environment would provide a vulnerability indicator.

Abstract representation. One of the sectors associated with the countermeasures portion of the model, *People Countermeasures*, is depicted in Figure 7. This sector can be seen to contain abstract qualitative entities such as "knowledge" and "satisfaction." These metrics can be established through surveys or perhaps through a point system applied to factors such as training and the state of the job market.

Relationships among variables in any SD simulation are cumulative, may run across sectors, and are indicated through the arrows in a diagram. A good example is presented by following a thread from Figure 7 down into Figure 8 below. In the above diagram of the *People Countermeasures* sector, an arrow emanates directly down from *Security Administrator Knowledge (SAK)* into the *Intrusion-Detection Analysis (IDA)* variable under *Process Countermeasures*. *Intrusion-Detection Analysis* is also affected by *IDS Knowledge Gain (IKG)* in the *Technology Countermeasures* sector.

Therefore: IDA = SKA * IKG; IDA   100

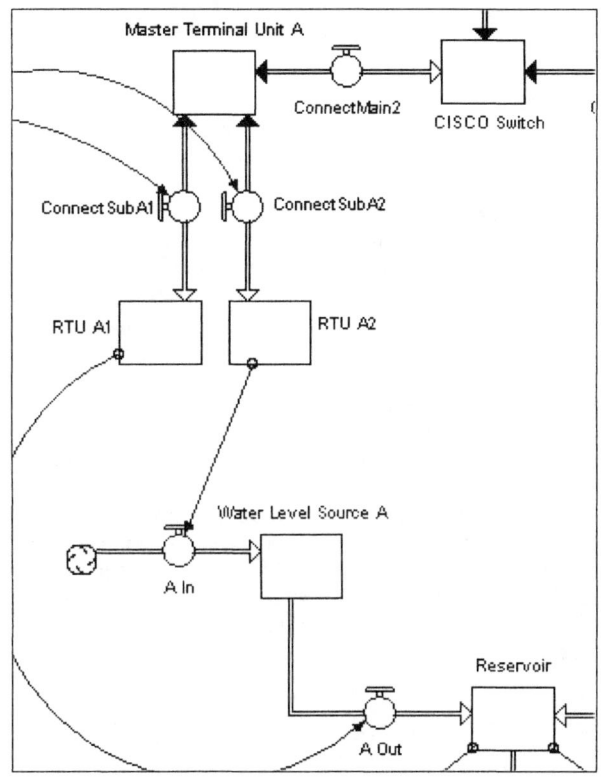

Figure 6. SCADA and physical connectivity.

Stated as prose, the intrusion-detection capability of an organization is affected by both installing IDS technology and by the educated ability of the information security staff to use it.

The *Technology Countermeasures* sector in Figure 8 presents an interesting study. The effectiveness of firewall technology is influenced by cost, by obsolescence, and by the need for programming. The ability to program a firewall is influenced by configuration management efforts, which are in turn influenced by the knowledge of information security personnel.

A Risk Metric

A high-level goal for users of this model is to minimize residual risk in their organization. *Residual risk* is that portion of risk that remains after security measures have been applied. It is the level of danger that a person or organization assumes and is willing to accept in operating a system. Figure 9 portrays residual risk stock in the *Risk Sector*. As portrayed, residual risk is increased by rising insider and outsider threats multiplied by vulnerabilities arising from the *Assets Sector* (not shown). It is decreased by applying appropriate countermeasures indicated in Figures 7 and 8; these include *people, process,* and *technology defense.* Ideally, residual risk is zero. It is up to the decisionmaker to determine how the factors affecting risk should be controlled. The model provides a sensitivity analysis ("what-if") framework for making this determination.

Ultimately, the organization needs to be concerned with the overall effectiveness of countermeasures against threats and vulnerabilities. While the complete elimination of risk is not possible, the manager should attempt to lower the mark as much as possible, yet remain within the economic, political, cultural, and technological constraints imposed upon him/her. For given resources, there will be multiple possible solutions lying on the Pareto optimal frontier.

A more complex variation of this model would be to match countermeasures, one for one, against vulnerabilities. Much of the knowledge and insights gained from a model such as this comes from the modeling effort. Organization personnel gain insights into leverage points by capturing the complexity of this large network of interrelating factors.

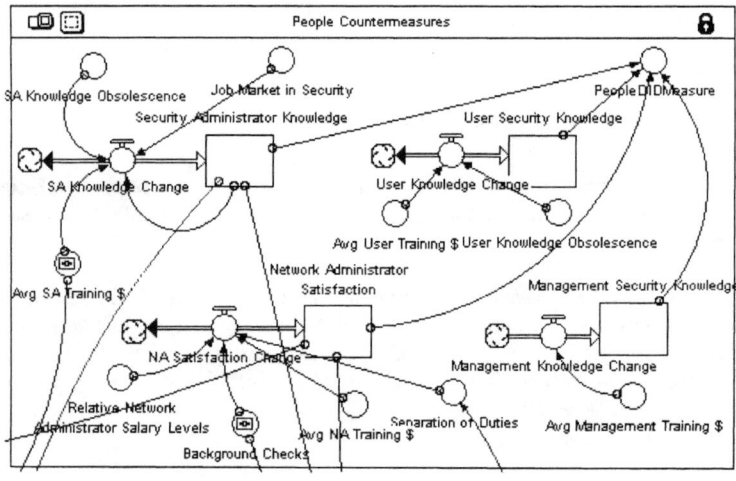

Figure 7. *People countermeasures* sector.

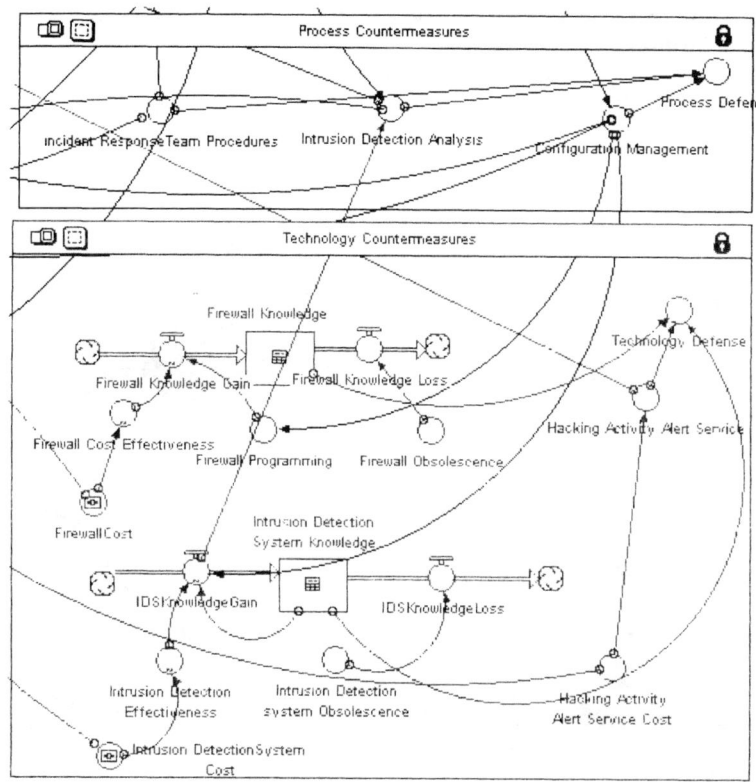

Figure 8. *Process* and *technology countermeasure* sectors.

Summary

The movement of infrastructure organizations toward greater standardization is bringing increasing levels of threats and vulnerabilities. The importance of timely reactions to attacks is particularly important in this environment. Countermeasures are available, but the challenge for infrastructure organizations is to best apply these with the limited resources they have available. Better methods are needed for justifying the added protection.

A system dynamics framework has been presented which is capable of moving beyond traditional risk assessment models. This system-focused framework provides the ability to focus on the issues of time, level of detail representation, and synthesis in information security. Details of a model were provided with an associated explanation of the modeling elements of stocks, flows, and causal effects. Further study is needed to validate the effectiveness of this proposed model against more traditional methods, such as cost/benefit using annual loss expectancy.

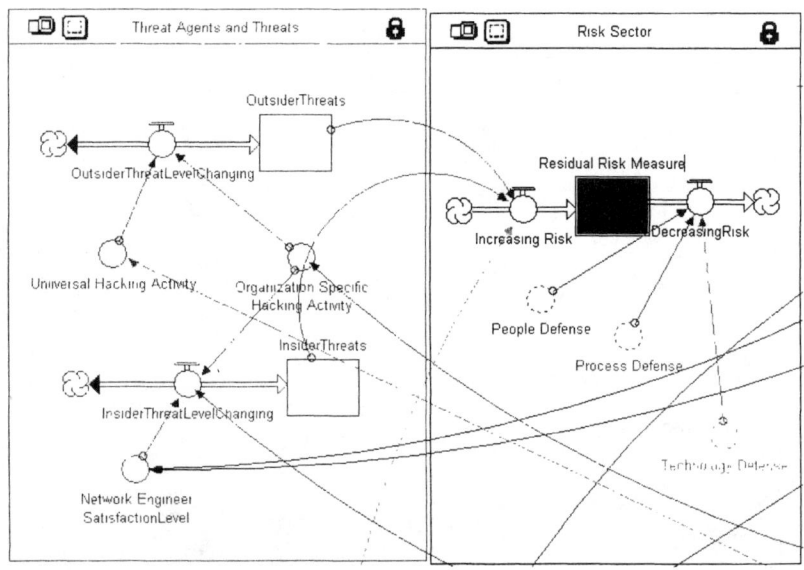

Figure 9. Residual risk.

References

AMSA (2002). *Asset Based Vulnerability Checklist.* Association of Metropolitan Sewerage Agencies, Washington, DC.

CCIMB (1999). *Common Criteria for Information Technology Security Evaluation: Part 1.* Introduction and General Model, version 2.1, Common Criteria Information Management Board, 99-031, August.

Ezell, B., J. Farr, and I. Wiese (2000). "Infrastructure risk analysis model." *Journal of Infrastructure Systems,* September.

Forrester, J. (1961). *Industrial Dynamics.* Williston, VT: Pegasus Communications.

Haimes, Y.Y. (1998). *Risk Modeling, Assessment, and Management.* New York: John Wiley and Sons.

Longstaff, T., C. Chittister, R. Pethia, and Y.Y. Haimes (2000). "Are we forgetting the risks of information technology?" *IEEE Computer,* December.

McClure, S., J. Scambray, and G. Kurtz (2001). *Hacking Exposed.* Berkeley, CA: Osbourne McGraw-Hill.

NIST (2002). *Guideline on Network Security Testing.* National Institute for Science and Technology, Publication 800-42.

SANDIA (2002). *Risk Assessment Methodology for Dams.* Sandia National Laboratory, http://tswg.gov/tswg/ip/RAMD_TB.htm.

USDOE (2002). *21 Steps to Improve Cyber Security of SCADA Networks.* US Department of Energy, Office of Energy Assurance, September.

Wenger, A., J. Metzger, and M. Dunn, (Eds.) (2001). *International Critical Information Infrastructure Protection Handbook,* Appendix A1, Glossary of Terms. Produced by Workshop on Critical Infrastructure Protection in Europe. Zurich, Switzerland, November 8-10. http://www.isn.ethz.ch/crn/extended/workshop_zh/ebp_cip_handbook.pdf.

The views expressed are those of the author and do not reflect the official policy or position of the National Defense University, the Department of Defense, or the US government.

Quantifying and Communicating Model Uncertainty for Decisionmaking in the Everglades

Daniel P. Loucks[1]

Abstract

One of this nation's largest ecosystem restoration projects is currently taking place in South Florida. The Comprehensive Everglades Restoration Project is a complex and expensive one, requiring a sequence of decisions over the next three decades in an attempt to preserve and enhance what remains of the Greater Everglades Region of South Florida. There is considerable uncertainty with respect to how this ecosystem functions and what decisions will be most effective in restoring it. This paper outlines some of the issues associated with estimating, quantifying, and communicating these uncertainties to various stakeholders, and to those who are responsible for decisionmaking.

Introduction

The Comprehensive Everglades Restoration Plan (CERP) involves numerous projects to be implemented over the next 30 years. This effort will also include monitoring the response of the system to projects as they are implemented. It is the responsibility of the Restoration Coordination and Verification (RECOVER) team to evaluate the system-wide performance of each project and the overall performance of the plan as it is carried out [SFWMD and USACE, 2003]. Depending on the results of these performance evaluations, the plan may be modified and refined.

Key to refining and implementing the CERP is evaluating the various performance measures derived from model output. Model uncertainty, originating from input uncertainty, parameter uncertainty, model structure uncertainty, and algorithmic (numerical) uncertainty, is translated into uncertainty in the results—the performance measures. In addition, there is uncertainty as to whether the specific performance measures used to characterize the overall system performance actually capture that performance.

---

[1]Professor, Civil and Environmental Engineering, Cornell University, Ithaca, NY 14853; 607-255-4896; DPL3@cornell.edu.

There is a need to quantify uncertainty for the specific performance measures used in RECOVER evaluations. Once this uncertainty is quantified, the RECOVER team responsible for evaluating model alternatives can take this uncertainty into account as they make decisions on the relative merits of each alternative. They will be seeking alternatives that have the highest probability of achieving project goals while minimizing the risk of undesired outcomes. This involves tradeoffs among target performance measures, their reliabilities (probability of not being met), and their costs. These tradeoffs are illustrated in Figure 1.

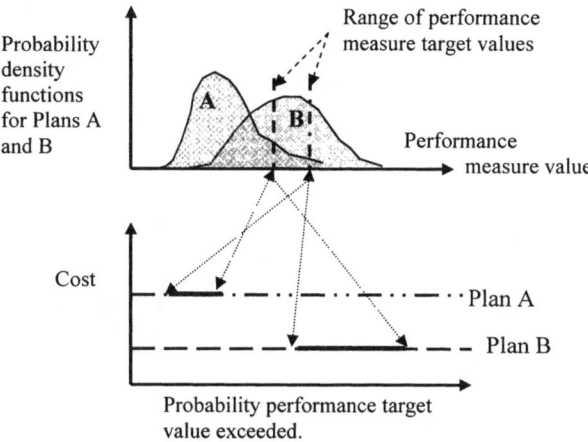

Figure 1. Tradeoffs among performance measure target values, the probability that the target values will not be met, and cost.

In this illustration the performance measure is a maximum desired value (e.g., the phosphorus concentration in a region of the Everglades). Values in excess of the target are considered failures.

What is Uncertain?

The Comprehensive Everglades Restoration Plan (CERP) is highly dependent on the results of dynamic regional simulation models. A main objective of CERP is to restore the ecosystem. The dynamic ecological models ELM [Fitz et al. 2002] and ATLSS [DeAngelis and Gross 2002] used to predict the responses of selected ecosystem variables to water management policies are dependent on the output of the hydrologic models (e.g., SFWMM and NSM [Tarboton et al. 2003]). Even though both the hydrologic and ecosystem models, and the improved models that may eventually replace some of them, are relatively complex and sophisticated compared to what actually takes place in the Everglades, like all such models, they are simplifications of reality. Hence their predictive abilities are not perfect. What is observed and what is predicted likely will not be the same. Even given the same boundary conditions—data that the models require as inputs—the observed results may not be the same. Hence the system response to any inputs, however measured, is best characterized by probability distributions. Given our current and foreseeable knowledge of how the Everglades works, it is unlikely that this uncertainty can be eliminated. Instead, we need to attempt to identify it and factor it into decisionmaking.

Each of these models used by CERP is deterministic. Given specific inputs, the outputs are always going to be the same each time those inputs are simulated. With specified inputs to any simulation model, if the predicted output does not agree with the observed value, as shown in Figure 2, this could result from measurement errors in both input and observed values. It could also result from errors in the model parameter values, the model structure, or the algorithm used to solve the model. None of these possible errors necessarily imply that uncertainty exists.

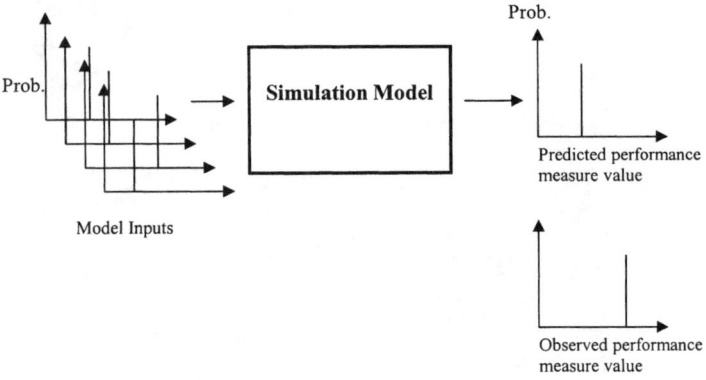

Figure 2. A deterministic system and a simulation model of that system needing calibration or modification in its structure.

There is no uncertainty, only parameter value or model structure errors to be identified and corrected in the above figure.

Next, consider the same deterministic simulation model, but now at least some of the inputs are random, i.e., not predictable. Clearly the performance measure outputs will also be unpredictable. Both the inputs and outputs are defined by probability distributions. If the uncertainty in the output is due only to the uncertainty in the input, the situation is similar to that shown in Figure 2. If the distribution of performance measure output values does not correspond to the distribution of observed performance measure values, then calibration of the model parameter values or modification of the model structure may be needed.

If a parameter calibration exercise finds the "best" values of the parameters to be outside reasonable ranges of values based on scientific knowledge, then the model structure or algorithm might be in error. Assume that the algorithms used to solve each of the CERP models are correct, and as is the case, observed measurements of system performance vary for the same model inputs, as shown in Figure 3. Then it can be assumed that the model structure does not capture all the ongoing processes that impact the value of the performance measures. This is often the case when relatively simple and low-resolution models are used to estimate the hydrological and ecological impacts of water and land management policies.

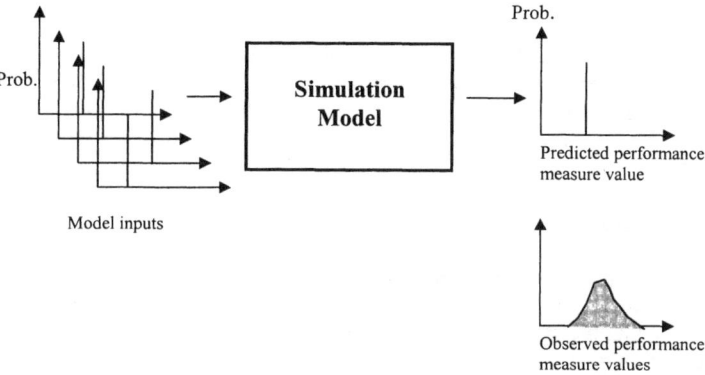

Figure 3. A deterministic simulation model of a *random* or *stochastic* system.

To account for the variability in observed results even given the same input values, the model's parameter values may need to vary over distributions of values, and/or the model structure may need modification along with additional model inputs.

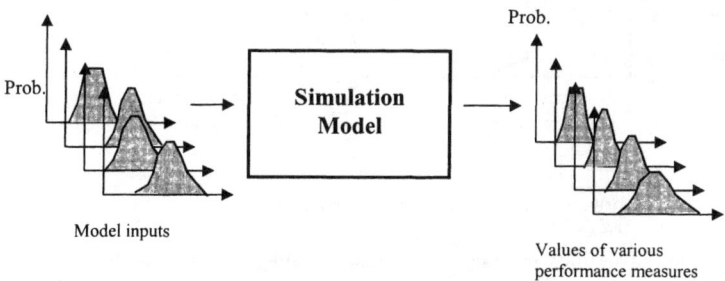

Figure 4. Simulating variable inputs to obtain distributions of predicted performance values that match the distributions of observed performance values.

Even if more detailed models requiring more input data and more parameter values were to be developed, capturing all the processes occurring in a system as complex as the Everglades is not likely to happen, nor is it likely to be cost-effective. Hence, those involved in CERP and RECOVER will have to live with what is commonly referred to as uncertainty. The issues addressed in this paper are how to estimate this variability or apparent uncertainty, and then how to utilize such information.

Uncertainty Analysis

What is needed is a way to predict the variability evident in the system shown in Figure 3. Instead of a fixed output vector for each fixed input vector, a distribution of outputs is needed for each performance measure based on fixed inputs (Figure 3) or a distribution of inputs (Figure 4). Furthermore, the model output distribution for each performance measure must "match" as much as possible the observed distribution of that performance measure.

General Concepts

An uncertainty analysis takes a set of stochastic input values (that can include parameter values), passes them through a model or transfer function, and then attempts to obtain the statistical distributions of the resulting outputs. As illustrated in Figure 5, the output distributions can be used to:
- describe the range of potential outputs of the system, and
- estimate the probability that the output will exceed a specific threshold or performance measure target value.

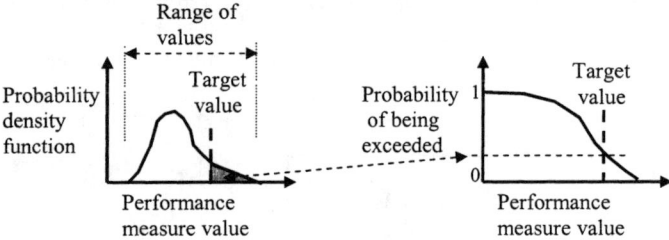

Figure 5. Distributions of performance measure values.

This distribution defines the range of potential values and the likelihood that any specified target value will be exceeded.

For CERP, a variety of performance measures must be supported by uncertainty analyses. In addition, these performance measures will have to be evaluated at a wide variety of locations.

What an Uncertainty Analysis Can Provide

The result of a numerically-based uncertainty analysis is a set of output values and the probability of occurrence for each value. These values are representative of the statistical distribution of the system outputs.

The most common uses for uncertainty analyses are to make general inferences such as the following:
- estimating the mean value and the standard deviation of the outputs,
- estimating the probability that the output will exceed a specific threshold (Figure 5),
- putting a confidence interval on a function of the outputs, e.g., the range of function values that are 90% likely to occur, and
- describing the complete range of potential outputs of the system (Figure 5).

Implicit in this approach are the assumptions that statistical distributions for the input values are correct, and that the model is a sufficiently realistic description of the processes taking place in the system. Neither of these assumptions is likely to be entirely correct.

The importance of the first assumption is easy to check by using different statistical distributions for the input parameters. If the outputs vary significantly compared to the original distributions, then the output is sensitive to the specification of the input distributions, and hence they must be defined with care. The assumption of a correct model choice can be supported in two ways. The first is to obtain consensus among the interested parties that the most correct or best model has been chosen. The second way is to develop and apply alternative models in a Monte Carlo analysis. Formal statistical approaches for handling multiple conceptual models are rarely used because they require developing an entire suite of plausible conceptual models.

An uncertainty analysis is not the same as a sensitivity analysis. An *uncertainty analysis* attempts to describe the entire set of possible outcomes, together with their associated probabilities of occurrence. A *sensitivity analysis* attempts to determine the relative change in model output values given small changes in model input values. A sensitivity analysis thus measures the change in the model in a localized region of the space of inputs. However, one can often use the same set of model runs for both uncertainty and sensitivity analyses [BIOMOVS 1993; Kann and Weyant 1999; Lal 1995].

Uncertainty of Performance Measure Targets

Another possible source of uncertainty is the selection of a target value for a performance measure. Just which target value is correct? When this is not clear, there are various ways of expressing the uncertainty associated with any target value. One such method is the use of fuzzy sets. Use of "grey" numbers or intervals instead of "white" or fixed target values is another. When some uncertainty or disagreement exists over the selection of the best target value for a particular performance measure, it seems to me that the most direct and transparent way is to subjectively assign confidence bands around some value that represents the range of possible target values. Then this subjective distribution can be factored into the tradeoff analysis, as outlined in Figure 6.

One of the challenges associated with defining and including the uncertainty associated with a target or threshold value for a performance measure is communicating just what the result of such an uncertainty analysis means. Referring to Figure 6, suppose that the target value represents some maximum limit of a pollutant concentration, say phosphorus, in the flow during a given period of time at a given site or region, and it is not certain just what that maximum limit should be. Subjectively defining the distribution of that maximum limit, and considering that uncertainty along with the uncertainty (probability of exceedance function) of pollutant concentrations—the performance measure—one can attach a confidence band to any probability of exceeding the maximum desired concentration value.

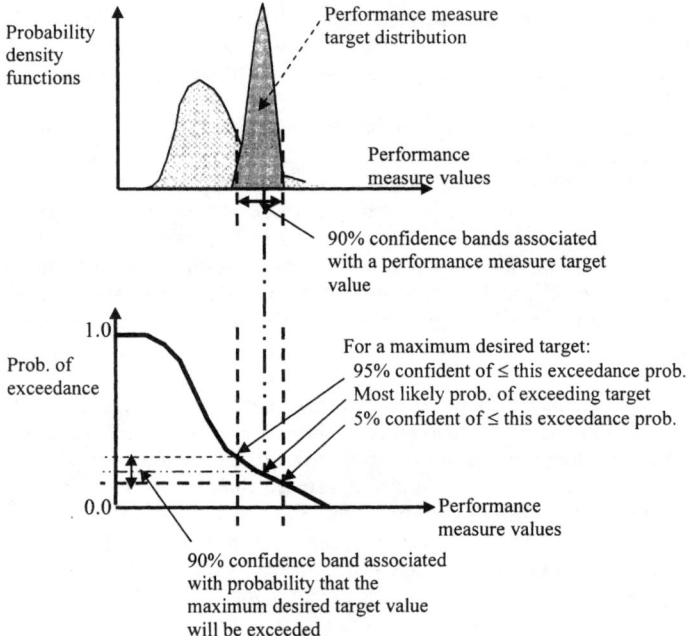

Figure 6. Probability distribution of performance measure values combined with probability distribution of performance measure target values.

As Figure 6 shows, combining the probability distribution of performance measure values with the probability distribution of performance measure target values helps to estimate the confidence one has in the probability of exceeding a maximum desired target value.

The 95% confidence probability of exceedance shown in Figure 6, say $P_{0.95}$, should be interpreted as: "We can be 95% confident that the probability of the maximum desired pollutant concentration being exceeded will be no greater than $P_{0.95}$." Likewise, we can only be 5% confident that the probability of exceeding the desired maximum concentration will be no greater than the lower $P_{0.05}$ value. Depending on whether the middle line through the subjective distribution of target values in Figure 6 represents the most likely or mean target value, the associated probability of exceedance is either the most likely, as indicated in Figure 6, or that for which we are only 50% confident.

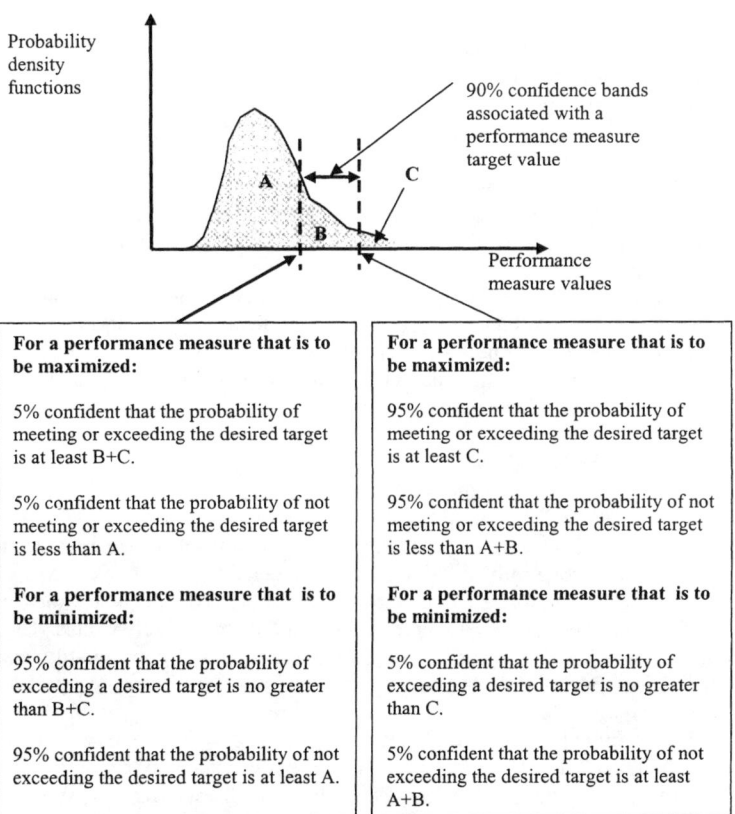

Figure 7. Interpreting confidence bounds with uncertain performance targets.

Figure 7 demonstrates that interpreting the results of combining performance measure probabilities with performance measure target probabilities depends on the type of

performance measure. The letters A, B, and C represent proportions of the probability density function of performance measure values (A+B+C = 1). Figure 7 attempts to show how to interpret the confidence bounds when the uncertain performance targets are:
- minimum acceptable levels that are to be maximized,
- maximum acceptable levels that are to be minimized, or
- optimum levels.

An example of a minimum acceptable target level might be the population of wading birds in an area of the Everglades. An example of a maximum acceptable target level might be, again, the phosphorus concentration in the flow in a specific region of the Everglades. An example of an optimum target level might be the depth of water most suitable for tree island development and maintenance during a period of the year.

For performance measure targets that are not expressed as minimum or maximum limits but are the "best" values, referring to Figure 7 we can state that we are 90% confident that the probability of achieving the desired target is no more than B. The 90% confidence-level probability of not achieving the desired target is at least A+C. The probability of the performance measure being too low is at least A, and the probability of it being too high is at least C, again at the 90% confidence level. As the confidence level decreases, the bandwidth decreases and at that lower confidence level, the probability of not meeting the target increases.

Clearly, there is uncertainty associated with each of these estimations, and this raises the question of just how valuable is the quantification of the uncertainty of each additional component of the plan evaluation process. Will plan evaluators and decisionmakers benefit from this additional information, and just how much additional uncertainty information is useful?

Communicating Uncertainty

There is no "best way" to communicate concepts of uncertainty to the public. The best way may well depend on what the public already knows about risk and the various types of probability distributions (e.g., density, cumulative, exceedance). Such knowledge would be based on objective and subjective data, and would include the distinction between mean or average values and the most likely values. Graphical representations of these ways of describing uncertainty considerably facilitate communication [Haimes 1998; Stern and Fineberg 1996; von Winterfeldt and Edwards 1986].

The National Research Council report, *Science and Judgment in Risk Assessment* [NRC 1994], addressed the extensive uncertainty and variability associated with estimating risk; it concluded that risk characterizations should not be reduced to a single number or even to a range of numbers intended to portray uncertainty. Instead, the report recommended that managers and the interested public should be given risk characterizations that are both qualitative and quantitative and both verbal and mathematical.

In some cases, communicating qualitative information about uncertainty to stakeholders and the public in general may be more effective than quantitative information. There are, of course, situations in which quantitative uncertainty analyses are likely to provide information that is useful in a decisionmaking process. How else can tradeoffs such as those illustrated in Figure 1 be identified? In addition, quantitative uncertainty analysis often can be used to improve qualitative information about uncertainty, even if the quantitative information is not communicated to the public.

When presenting the results of any modeling applied to the Everglades, it should not be too difficult for the public to realize that, whether or not such models explicitly included uncertainty, the model predictions will not necessarily correspond to what may be observed. There is no way to accurately predict any particular system-performance measure. First of all, the models used are relatively simple compared to reality. Secondly, many features of the Everglades, such as the weather, are variable, and this variability can be measured. Third, we just do not know enough about the processes that take place in the Everglades to eliminate the possibility of surprises. To account for variability, it is possible to identify and measure the range of possible outcomes and their probabilities or likelihoods of occurring. To account for unawareness is much more difficult. We have to expect that unforeseen events or outcomes will occur, and even if they could be identified in advance, there is no way to estimate their likelihoods.

We should acknowledge to the public the widespread confusion regarding the differences between variability and uncertainty. Variability does not change through further measurement or study, although better sampling can improve our knowledge of it. Uncertainty reflects gaps in information about scientifically observable phenomena. Uncertainty sometimes can be reduced through further measurement or study, and then by including this increased understanding within the simulation model.

Support for routine formal quantitative analysis of uncertainty is based on the desire to move away from using average values or point estimates that do not identify 1) the range of possible values or 2) the confidence that can be associated with any particular performance measure value. Providing a numerical range of possible risks that reflects uncertainty and variability is thought to allow more-informed and more-transparent decisions than are possible when only a single point estimate is generated. The level of detail with respect to uncertainty and risk that needs to be effectively communicated to decisionmakers (or plan evaluators) will likely differ and be more quantitative than the information concerning uncertainty desired by other stakeholders.

We must effectively communicate information about who or what is at risk or what might happen and just how severe and irreversible an adverse effect might be should a target value not be met. This is just as important as communicating the level of uncertainty and the confidence associated with such predictions. This may be qualitative information, but it is often critical to informed decisionmaking. The communication of risk and uncertainty is always complicated by the questions of how much information is enough and how best to present it. Feedback and communication between those receiving such information and those giving it can help identify what seems best for a particular audience.

Uncertainty characterizations must include information that is useful for all parties participating in a decisionmaking process involving tradeoffs between performance target values, probabilities of those targets not being met, and cost. Quantitative estimates of uncertainty are needed for this, but supplementary qualitative information on the nature of any possible adverse effects and how the uncertainty estimates were obtained is also likely to be useful.

Communicating a range or distribution of performance measure values reflecting uncertainty can be perplexing to nontechnical stakeholders who often want technical staff to tell them whether a plan will work or not. Such information should not be perplexing to those involved in RECOVER. Yet the more simple and understandable the analyses, the more useful they may be in RECOVER. If there is a choice between academically interesting but rather complex ways of quantifying uncertainty, and on the other hand, conducting research (e.g., adaptive management experiments) focused on reducing important sources of uncertainty, spending money on reducing uncertainty would seem preferable to spending it on ways of calculating and describing it better.

In spite of some considerable efforts by those involved in risk assessment and management, we know very little about how to ensure effective risk communication that gains the confidence of stakeholders, incorporates their views and knowledge, and influences favorably the acceptability of risk assessments and risk management decisions. This suggests that RECOVER will need to consider adopting comprehensive communication programs; these would help stakeholders as well as decisionmakers to better understand the implications of 1) working or living with the uncertainties inherent in trying to restore an ecosystem as complex as the Everglades, and 2) having to make tradeoffs between targets, reliabilities, and costs.

Studies of the differences between technical and nontechnical perceptions of risk and uncertainty have identified many of the factors that contribute to misunderstandings and resentment when events do not turn out to be as expected. The discussion here is not comprehensive; rather, it is intended to indicate the importance of effective communication and the potential for mistakes and misunderstandings.

Risk and uncertainty communication is a two-way street—it means both learning and teaching. Communicators dealing with uncertainty should learn about the concerns and values of their audience, their relevant knowledge, and their experience with uncertainty issues. Stakeholders' knowledge of sources and reasons for uncertainty analyses needs to be incorporated into assessment and management decisions. By listening, communicators can craft risk messages that better reflect the perspectives, technical knowledge, and concerns of the audience.

Effective communication must begin before important decisions have been made. It can be facilitated in communities by citizen advisory panels. These can give planners and decisionmakers a better understanding of the questions and concerns of the community and an opportunity to test its effectiveness in communicating concepts and specific issues regarding uncertainty.

One approach to make uncertainty more meaningful is to make risk comparisons. For example, ten parts per billion, the target maximum phosphorus concentration in the Everglades, is equivalent to 10 seconds in over 31 years. If this is an average daily concentration target that is to be satisfied "99 percent" of the time, it is equivalent to an expected violation of less than one day every three months.

Even more difficult to communicate is the fact that a 1-in-100 probability of exceedance estimate is not an estimate of actual probability that an exceedance event will occur, but a statistical upper bound on the likelihood that it could happen. The actual risk may be much lower.

Many perceive the reduction of risk by an order of magnitude as though it were a linear reduction. A better way to illustrate orders of magnitude of risk reduction is shown in Figure 8, in which a bar graph depicts better than words that a reduction in risk from 1-in-1,000 (10^{-3}) to 1-in-10,000 (10^{-4}) is a reduction of 90% and that a further reduction to 1-in-100,000 (10^{-5}) is a reduction 10-fold less than the first reduction of 90%. The percent of the risk that is reduced by whatever measures is a much easier concept to communicate than reductions expressed in terms of estimated absolute risk levels, such as 10^{-5}.

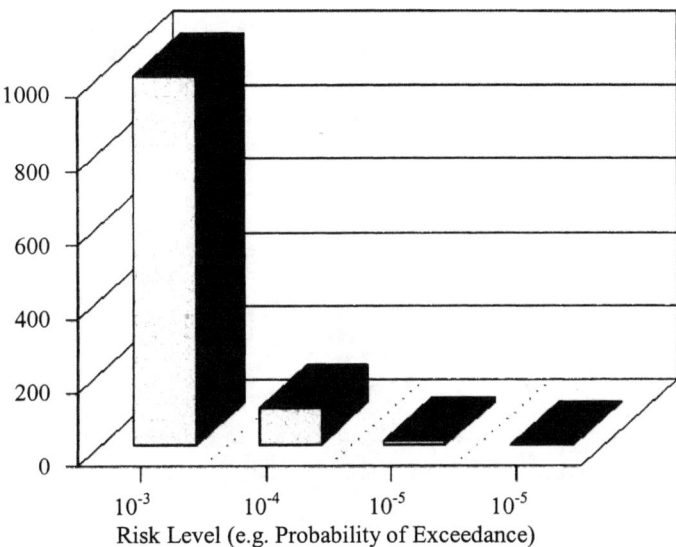

Figure 8. Bar graph illustrating orders of magnitude of risk reduction.

Figure 8 clearly shows that reducing risk by orders of magnitude is not equivalent to linear reductions.

Alternative ways of displaying uncertainty have been outlined by many [e.g., Anderson 1998; Covello and Merkhofer 1993; DOE 1998; Kann and Weyant 1999; Kelly and Roy-Harrison 1998; Reckhow 1994; Suter 1993; USEPA 1992]. Figures 9 and 10 combine some of them to illustrate time-series data and their trends and ranges.

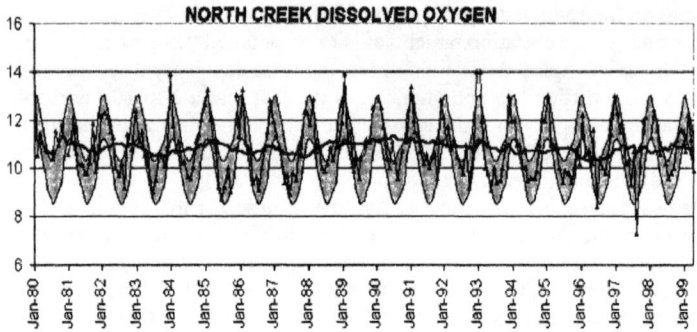

Figure 9. Alternate comparison graph for displaying uncertainty.

Figure 9 shows the actual annual and monthly mean values of dissolved oxygen concentrations in a stream together with the ranges of the monthly average values occurring 90% of the time [King County 2003].

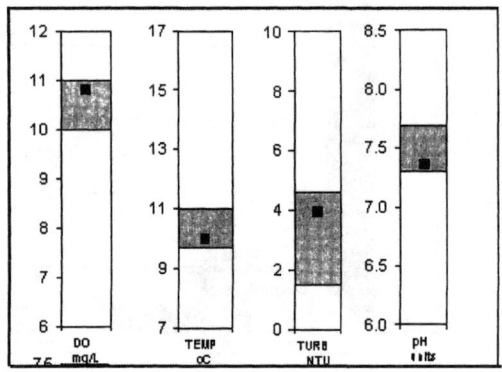

Figure 10. Values of selected water quality variables [King County 2003]. The gray bands identify the mid-50-percent range of values.

Tables 3 and 4 also include indicators of distributions associated with measured (or modeled) data. These plots and tables were taken from the web pages of King County [King County 2003].

Table 3. Comparison of sediment metal concentrations (mg/kg dry weight) with state freshwater sediment thresholds.

Metal	Sediment Threshold	Mean	Min-Max
Arsenic	None	< 5.5	< Det-9.2
Silver	None	< Det	< Det
Mercury	2	< Det	< Det
Cadmium	10	< 0.23	< Det-0.39
Copper	110	7.0	5.7-8.7
Lead	250	3.1	3.1

Table 4 (next page) summarizes the water quality characteristics of the stream under baseflow conditions and whether state or federal water quality criteria standards were met. Table headings are explained below.

Heading	Description
n =	Number of measurements between 1979 and 1999. Most sites have been sampled monthly.
Mean	Arithmetic average. (If the parameter has occasional extreme values, the mean can become misleading.)
Minimum	Lowest value measured at the site.
Maximum	Highest value measured at the site.
Median (+/- 1 s.d.)	Range into which the middle 70% of the values fell. This has the advantage of not being distorted by a few extreme values over the period of record. It is derived by first sorting the values in ascending or descending order and determining the middle value—the median. Second, the standard deviation (s.d.) is calculated to determine how variable the data are. Finally, the endpoints of the range are calculated—the median minus 1 standard deviation is the low end and the median plus 1 standard deviation is the high end.
# Non-standard	Some parameters have water quality criteria assigned to them. The values in this column are the number of measurements that did not meet those criteria.
% Non-standard	The value described above expressed as a percentage.

Table 4. A summary of water quality characteristics and measurement results [King County 2003].

Mouth of North Creek (0474)

	n=	Mean	Minimum	Maximun	Median +/- 1 S.D	# Non-standard	%Non-standard
FLOW (CFS)	175	38.39	1.3	300	10.8-58.0	N/A	N/A
D.O. (MG/L)	231	10.84	7.3	14.02	9.7-12.0	16	7.10
TEMPERATURE (°C)	245	9.98	0.1	20	4.7-18.4	17	8.94
TURBIDITY (NTU)	234	3.89	0.2	30	1.8-5.7	15	6.41
pH (UNITS)	231	7.35	6.09	8.4	7.2-7.8	1	0.43
CONDUCTIVITY (µMHO)	234	152.6	61	570	115-189	N/A	N/A
TSS (MG/L)	235	7.72	1.25	97.14	3.2-11.4	N/A	N/A
ORTHO-P (MG/L)	233	0.0506	0.007	0.23	.029-.071	N/A	N/A
TOTAL-P (MG/L)	233	0.0886	0.0394	0.373	.060-.115	N/A	N/A
AMMONIA (MG/L)	192	0.0375	0.001	0.112	.011-.069	N/A	N/A
NITRATE (MG/L)	232	0.8699	0.001	1.89	.487-1.374	N/A	N/A
TOTAL-N (MG/L)	76	1.2898	0.935	2.07	1.038-1.540	N/A	N/A
ENTEROCUCCUS (CFU/100ml)	127	100	10	2600	28-293	64	50.39
FECAL COLIFORM (CFU/100ml)	235	248	0	7500	71-737	172	73.19

Uncertainty and Decisionmaking

Consider the tradeoffs that need to be made as illustrated in Figure 1. That figure is repeated here as Figure 11, but now assume that the range of target values represents the 90%-confidence range associated with an uncertain single performance measure target value that is a maximum desired upper limit (e.g., phosphorus concentration).

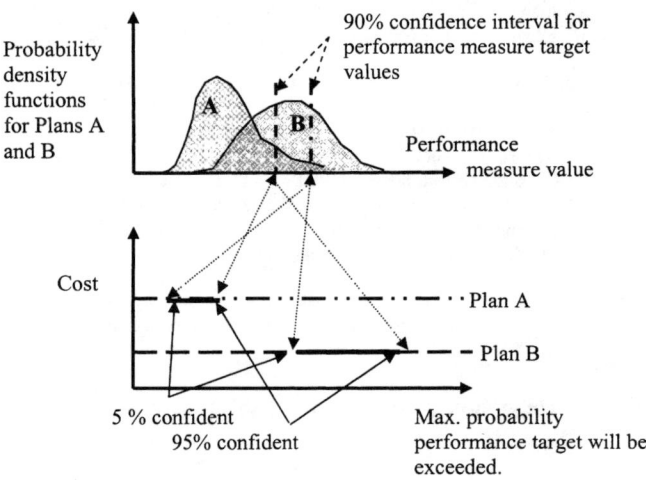

Figure 11. Two plans showing ranges of maximum probabilities.

Figure 11 shows two ranges, depending on one's confidence, of the maximum probability that an uncertain desired upper limit performance target value will be exceeded. The 95% confidence levels are associated with the higher probabilities, and the 5% confidence levels are associated with the more desirable lower probabilities of exceeding the upper limit target.

In the case shown above, the tradeoff is clearly between cost and reliability, since no matter what the confidence, Plan A is preferred to Plan B with respect to reliability, but Plan A is more expensive than Plan B. The tradeoff is between reliabilities and costs only.

Consider however a third plan, Plan C, as shown in Figure 12.

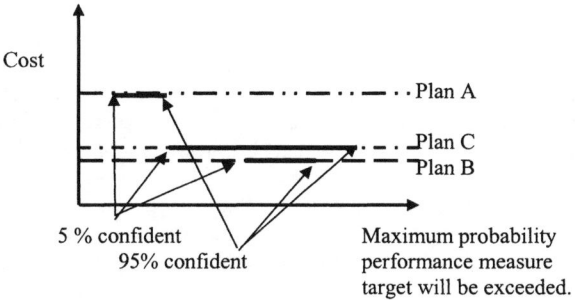

Figure 12. Tradeoffs among cost, reliabilities, and the confidence level of those reliabilities.

The relative ranking of plans with respect to the probability of exceeding the desired (maximum limit) target may depend on the confidence given to that probability.

This situation adds to the complexity of making appropriate tradeoffs, as now there are three criteria: cost, probability of exceedance (reliability), and the confidence in those reliabilities or probabilities. Add to this the fact that there will be multiple performance measure targets, each expressed in terms of their maximum probabilities of exceedance and the confidence in those probabilities.

If the plan evaluation process has difficulty handling all this, it may indicate the need to focus the uncertainty analysis effort on just what is deemed important, doable, and beneficial. When the alternatives have been narrowed down to only a few plans that appear to be best, a more complete uncertainty analysis can be performed.

From uncertainty analyses we can gain an appreciation of the magnitude of the uncertainty associated with model predictions and use that information when evaluating alternative CERP management plans or projects. In the specific area of water and ecosystem management, most science-based decisions have to deal with incomplete or inaccurate science. For example, everyone involved in CERP knows that models cannot predict with precision the effect of water management changes on water quantity and quality distribution over space and time, let alone their impacts on multiple ecological performance measures. Yet even with this uncertainty, planners and decisionmakers have to rely on those model predictions to guide decisions. Models are the best tools available, barring implementing an actual management policy and waiting a long time to see how it works.

Some Conclusions

Uncertainty analysis in support of environmental management is motivated by the fact that environmental management is usually uncertain, and sometimes highly so. The opportunity that is provided by uncertainty analysis should result in more informed decisionmaking. But it comes at a cost, and this cost should be considered along with the benefits of having uncertainty information.

There are many interesting research opportunities and needs with respect to carrying out an uncertainty analysis on a system as complex as the Everglades, as well as with the models being used to identify and evaluate alternative water management policies. Clearly there is uncertainty associated with any prediction of what might happen if any particular water management policy is implemented. Nevertheless, in any scientific study in support of uncertainty analyses for improved decisionmaking, the focus should be on the decisions being made or the objectives associated with the resource, and not on an academic model or on basic science. A predictive model (or more generally, a predictive scientific assessment) should be evaluated in terms of its use in addressing these decisions/objectives. This means that ideally, predicted endpoints should be decision-based and not tool-(model)-based. In reality, good endpoints will reflect a compromise between what is desirable to aid decisionmaking and what is feasible for scientific assessment.

Acknowledgment

This paper is a condensed version of a report by Lall et al. [2002].

References

Anderson, J.L. (1998). "Embracing uncertainty: the interface of Bayesian statistics and cognitive psychology." *Conservation Ecology* [online] 2(1), 2. Available from the internet, http://www.consecol.org/vol2/iss1/art2.

BIOMOVS (1993). *Biospheric Model Validation Study, Phase II, Technical Report No. 1, Guidelines for Uncertainty Analysis*. Swedish Radiation Protection Institute, Stockholm, Sweden.

Covello, V.T. and M.W. Merkhofer (1993). *Risk Assessment Methods*. London: Plenum Press.

DeAngelis, D. and L.J. Gross (2002). *ATLSS: Across Trophic Level System Simulation*. Institute for Environmental Modeling, University of Tennessee, Knoxville, TN, http://atlss.org/.

DOE (1998). *Screening Assessment and Requirements for a Comprehensive Assessment*. DOE/RL-96-16, Rev. 1, US Department of Energy, Richland, WA.

Fitz, H.C. et al. (2002) *The Everglades Landscape Model (ELM)*. South Florida Water Management District (SFWMD), West Palm Beach, FL, http://www.sfwmd.gov/org/wrp/elm/index.html.

Haimes, Y.Y. (1998). *Risk Modeling, Assessment, and Management*. New York: John Wiley & Sons, Inc.

Kann, A. and J.P. Weyant (1999). "A comparison of approaches for performing uncertainty analysis in integrated assessment models." *Journal of Environmental Management and Assessment*, 5(1), 29-46.

Kelly, E. and W. Roy-Harrison (1998). "A mathematical construct for ecological risk: a useful framework for assessments." *Human and Ecological Risk Assessment, An International Journal*, 4(2), 229-241.

King County (2003). Streams monitoring pages, Water and Land Resources Division, King County Department of Natural Resources, Seattle, WA, http://dnr.metrokc.gov/wlr/waterres/streams/creekindex.htm.

Lal, W. (1995). "Sensitivity and uncertainty analysis of a regional model for the natural system of South Florida." Draft report, South Florida Water Management District (SFWMD), West Palm Beach, FL, November.

Lall, U., D.L. Philips, K.H. Reckhow, and D.P. Loucks (2002). *Quantifying and Communicating Model Uncertainty for Decision Making in the Everglades*. USACE/SFWMD Model Uncertainty Workshop report, Central and Southern Florida Project, July.

NRC (1994). *Science and Judgment in Risk Assessment*. National Research Council, National Academy Press, Washington, DC.

Reckhow, K.H. (1994). "Water quality simulation modeling and uncertainty analysis for risk assessment and decision making." *Ecological Modelling*, 72(1), 20.

SFWMD and USACE (2003). *Everglades Restoration Plan*. South Florida Water Management District, West Palm Beach, FL, and US Army Corps of Engineers, Jacksonville, FL, http://www.evergladesplan.org/.

Stern, P.S. and H.V. Fineberg, (Eds.) (1996). *Understanding Risk: Informing Decision in a Democratic Society*. National Research Council (NRC) Committee on Risk Characterization, National Academy Press, Washington, DC.

Suter, G.W. II (1993). *Ecological Risk Assessment*. Boca Raton, FL: Lewis Publishers.

Tarboton, K. et al. (2003). *South Florida Water Management Model and Natural Systems Model*, South Florida Water Management District, West Palm Beach, FL, http://www.evergladesplan.org/pm/recover/mrt_model_review.cfm.

USEPA (1992). *Framework for Ecological Risk Assessment*. EPA/630/R-92/001, US Environmental Protection Agency, Risk Assessment Forum, Washington, DC.

von Winterfeldt, D. and W. Edwards (1986). *Decision Analysis and Behavioral Research*. Cambridge, UK: Cambridge University Press.

Optimal Allocation of Resources for Defense of Simple Series and Parallel Systems from Determined Adversaries

Vicki M. Bier[1] and Vinod Abhichandani[2]

Abstract

Managing the risks posed by an intelligent and adaptable adversary is different from many other types of risk management. Thus, risk management in this context is a problem of game theory rather than decision theory. In other words, the defender wishes to choose the optimal strategy for defending against an optimal attack, and vice versa. In this paper, we apply game theory to help in characterizing optimal defensive strategies against intentional attacks. The results yield useful insights.

Introduction

In the aftermath of the September 11th, 2001, attacks on the World Trade Center and the Pentagon (and the anthrax attacks in the United States), there is increased interest in strategies for protecting assets of value (including human life) against attacks by an intelligent and adaptable adversary. Even before that time, there were calls for greater attention to critical infrastructure protection, including computer security; see, e.g., President's Commission on Critical Infrastructure Protection [1997].

This is a fundamentally different challenge from protecting against "acts of nature" or "accidents." For example, an earthquake will not become stronger or "smarter" just because we have hardened our buildings to protect against it. In contrast, an intelligent and determined adversary is likely to adopt a different offensive strategy once we have put a particular set of protective measures in place. Therefore, good defensive strategies must consider the adversary's behavior.

---

[1]Professor, Dept. of Industrial Engineering, University of Wisconsin-Madison, 1513 University Avenue, Madison WI 53706; 608-262-2064; bier@engr.wisc.edu
[2]Graduate student, Dept. of Industrial Engineering, University of Wisconsin-Madison, 1513 University Avenue, Madison, WI 53706; 608-263-2687; vinod_abhi@yahoo.com

To illustrate, if we harden one point of entry into a system to make it essentially invulnerable, an adversary is likely simply to target an alternative point of entry instead. Thus, Schneier [2001] notes: "The defender has to defend against every possible attack. The attacker, on the other hand, only has to choose one attack, and he can concentrate his forces on that one attack."

The requirement to take into account the behavior of adversaries means that we must consider their goals and motivations when we select defensive strategies [Schudel and Wood 2000]. One of the most important characteristics may be whether an adversary is opportunistic or determined. Opportunistic adversaries may simply be looking for easy targets. In other words, they may be interested in a wide range of targets and shift to an easier one if their current target proves too difficult or costly to attack. Ordinary vandalism typically falls into this category; similarly, many computer hackers may not care which internet businesses they disrupt. In contrast, realistic levels of difficulty or cost will not necessarily deter a determined attacker. Examples of determined attackers might include military adversaries or terrorists, unethical business competitors, or even disgruntled ex-employees who are determined to damage a company's assets or reputation.

Defending against opportunistic adversaries is less difficult than defending against determined attackers. As in the old joke about the two campers who see a bear, in defending against opportunistic attackers you do not have to run faster than the bear; you just have to run faster than the other potential victim does! Thus, if an organization is convinced that most of its adversaries are opportunistic, it may be acceptable merely to adopt security practices that are better than the average for similar companies or systems. In this type of situation, the organization's defenses need only be strong enough to discourage possible attacks.

Defending something of value from a truly determined attacker is much more difficult. To illustrate some of the challenges involved, we will compare the optimal strategies for defending simple systems (configured either in series or in parallel) against attacks by knowledgeable and determined adversaries.

General Problem Formulation and Notation

Consider a system made up of multiple components. Redundant components that provide alternative means to perform the same task are functionally parallel to each other. Similarly, components that are all necessary to perform a single task are functionally in series. Of course, real-world systems may consist of a large number of components in complex combinations of parallel and series configurations. As a building block toward understanding optimal strategies for investing in the security of such systems, we discuss simple systems with n components, either in parallel or in series with each other, subjected to limited attacks by an intelligent adversary.

To derive a better understanding of the security of such systems, we first develop mathematical formulations for systems with only two components, corresponding to various possible assumptions regarding attacker and defender knowledge, constraints, and objectives. We then extend the results to systems with more than two components. We assume that the attacker wishes to maximize the probability of success for an attack on the system. In addition, we assume that attacks against different components succeed or fail independently of each other. The assumption of independence is obviously somewhat restrictive; it is likely to be more nearly satisfied in systems that have already invested in basic reliability and security measures such as spatial separation and functional diversity.

We summarize below some mathematical notations and symbols used throughout the paper:

C: The total budget constraint (where applicable).

C_i: The investment allocated to defend component i.

$P(C_1, C_2, ..., C_n)$: The probability of success of an attack against the system, as a function of the defensive investments in all components.

$P_i(C_i)$: The probability of success of an attack against component i, as a function of the resources C_i expended to strengthen that component. The functions $P_i(C_i)$ are deemed to be convex, decreasing, strictly positive, continuous, twice differentiable, and invertible (with differentiable inverses).

α: The expected dollar value of reducing the probability of success of an attack against the network from one to zero; e.g., the dollar value of the system if operational, multiplied by the probability of an attack against the system.

$F(C_1, C_2, ..., C_n)$: The objective function that the defender wishes to minimize in any given case. In the case where the defender has a budget constraint, this will simply be the probability of success of an optimal attack against the system. In the case with no budget constraint, the objective function will be a weighted sum of the probability of success of an optimal attack against the system and the defensive investments made by the defender, with the weight given by the parameter α.

Components in Parallel

To begin, consider a system of two components in parallel. Since the components are redundant, either one is sufficient to ensure successful operation of the system. Therefore, attackers with the goal of disabling the network would have no choice but to attack and disable both components, even though they may have a choice about which component to target first. We assume that the defender is concerned about the probability that an attack will succeed, but not the time at which the attack occurs, so it does not matter whether the two attacks occur simultaneously or sequentially.

If the defender spends C_1 on protection of component 1 and C_2 on protection of component 2, then under the assumption of independence, the resulting probability of an attack on the network succeeding will be given by $P(C_1, C_2) = P_1(C_1) P_2(C_2)$. We now develop mathematical models for optimal defense of this system for two possible cases, depending on whether the defender has a cost constraint.

Constrained case. Here, we assume that the defender has limited resources available for enhancing the system defenses, and wishes to maximize the system security subject to the given budget constraint. Therefore, the objective function to be minimized by the defender is simply the total probability of an attack on the system succeeding; that is,

$$F(C_1, C_2) = P(C_1, C_2) = P_1(C_1) P_2(C_2) \qquad (1)$$

subject to the constraints $C_1 + C_2 \leq C$ and $C_i \geq 0$.

It is easy to show that the objective function is monotonically decreasing. Therefore, the constrained minimum must occur along the line $C_1 + C_2 = C$; thus, at the minimum, we will have $C_2 = C - C_1$. Substituting this into the objective function, differentiating with respect to C_1, and equating the result to zero, we get:

$$\frac{\partial F(C_1, C - C_1)}{\partial C_1} = P_1'(C_1) P_2(C - C_1) - P_1(C_1) P_2'(C - C_1) = 0$$

Thus, if a local optimum exists for $0 < C_1 < C$, it must satisfy:

$$\frac{P_1'(C_1)}{P_1(C_1)} = \frac{P_2'(C - C_1)}{P_2(C - C_1)}$$

To determine whether a point satisfying this condition is a minimum, a maximum, or an inflection point, we take the second derivative of the objective function:

$$\frac{\partial^2 F(C_1, C - C_1)}{\partial C_1^2} = P_1''(C_1) P_2(C - C_1) + P_1(C_1) P_2''(C - C_1) - 2 P_1'(C_1) P_2'(C - C_1)$$

A point satisfying the first-order condition will be a local minimum if this second derivative is positive, a local maximum (along the constraint $C_1 + C_2 = C$) if the second derivative is negative, and inconclusive if the second derivative equals zero.

Note that the global minimum also can occur at a point where one of the C_i equals zero and the other is equal to C, even if this point does not satisfy the first-order condition given above. For example, one component may be so much more costly to defend than the other that the optimal defense of the system involves a zero expenditure on defense of the more costly component. An optimum with no investment in one component can occur under several conditions. First, if there is only one critical point along the constraint $C_1 + C_2 = C$, it may be a local maximum or an inflection point (when viewed as a one-dimensional function of C_1). Second, there may be more than one critical point along the constraint $C_1 + C_2 = C$. Finally, there may be no critical point along the constraint $C_1 + C_2 = C$.

Special cases.
1. Let $P_i(C_i) = a_i e^{-b C_i}$. Here, the first-order condition for optimality given above will be satisfied for all values of C_1 along the line $C_1 + C_2 = C$, so any point along this constraint will be a global minimum.

2. Let the $P_i(C_i)$ be log-convex (which implies, roughly speaking, that the probability of success of an attack against component i decreases faster than exponentially in the level of defensive resources invested). In this case, the

defender's objective function will be convex. Therefore, if a feasible point satisfying the first-order optimality condition exists, it will be the unique global minimum. However, if a feasible point satisfying the first-order optimality condition does not exist, the global minimum will occur where C_1 equals either zero or C, as before.

3. Let the $P_i(C_i)$ be identical, so that $P_1(C_1) = P_2(C_1)$ for all C_1. Substituting this into the objective function from Equation (1), we get a new objective function, $F(C_1, C - C_1) = P_1(C_1) P_2(C - C_1) = P_1(C_1) P_1(C - C_1)$. To find the minimum, we differentiate with respect to C_1 and then equate the result to zero, giving $F'(C_1, C - C_1) = P_1'(C_1) P_1(C - C_1) - P_1(C_1) P_1'(C - C_1) = 0$, or:

$$\frac{P_1'(C - C_1)}{P_1(C - C_1)} = \frac{P_1'(C_1)}{P_1(C_1)}$$

As before, there can be multiple local optima, although now those optima must be symmetric. One critical point occurs at $C - C_1 = C_1$, or $C_1 = C/2$. However, this point need not be the global minimum, unless the objective function is convex (e.g., if the P_i are log-convex). For example, it may be more cost-effective to invest all defensive resources in a single component, in which case symmetric global minima would occur at $C_1 = 0$ and $C_1 = C$.

Extension to systems with more than two components. For a system with n components in parallel, the attacker must disable all components. Therefore, the probability of an attack succeeding against a system with n functionally redundant components is given by $P(C_1, C_2, ..., C_n) = P_1(C_1) P_2(C_2) ... P_n(C_n)$. To find the constrained minimum of this function, we again use the fact that the objective function is monotonically decreasing. Therefore, the constrained minimum must occur along the constraint $C_1 + C_2 + ... + C_n = C$; i.e., at a point satisfying:

$$C_1 = C - (C_2 + ... + C_n) \tag{2}$$

Substituting Equation (2) into the objective function yields a revised objective function given by $F(C_1, C_2, ..., C_n) = P_1[C - (C_2 + ... + C_n)] P_2(C_2) ... P_n(C_n)$. To obtain the optimum of this function, we differentiate with respect to C_i for $i = 2, 3...n$, and equate the results to zero, yielding:

$$\frac{P_1'[C - (C_2 + ... + C_n)]}{P_1[C - (C_2 + ... + C_n)]} = \frac{P_i'(C_i)}{P_i(C_i)}$$

and therefore in general:

$$\frac{P_i'(C_i)}{P_i(C_i)} = \frac{P_j'(C_j)}{P_j(C_j)} \tag{3}$$

Thus, if the global optimum occurs at a point where both C_i and C_j are non-zero, then at optimality, the percentage decrease in P_i caused by spending more on component i will equal the percentage increase in P_j caused by spending less on component j. As before, the optimum may involve spending no resources on one or more components (e.g., if they are much more costly to defend than others), in which case the first-order conditions given in Equation (3) may not be satisfied for pairs of components that do not both have positive defensive investments at optimality.

For example, consider a system made up of three components. For the reasons discussed above, the optimum must lie at $C_1 + C_2 + C_3 = C$. Figure 1 below graphically depicts the possible values of the C_i at optimality. If the global optimum occurs at an interior point of the triangle described by the points $C_1 = C$, $C_2 = C$, and $C_3 = C$, the first-order conditions given in Equation (3) will be satisfied for all pairs of the variables. If the global optimum occurs along an edge of the triangle (with one of the C_i zero), then only the two non-zero variables must satisfy Equation (3). Finally, the global optimum may occur at a vertex of the triangle (where one of the C_i equals C), in which case the first-order conditions in Equation (3) may not be satisfied for any of the three variables.

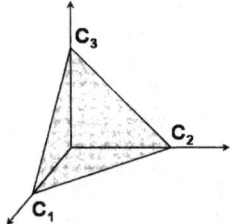

Figure 1. Possible variable values at optimality.

Interpretation of results. $P_i'(C_i)$ is the rate of change of the probability of success of an attack on component i. Dividing the rate of change by $P_i(C_i)$ essentially converts an absolute rate of change to a fractional rate of change. Therefore, the first-order conditions in Equation (3) essentially mean that if C_i and C_j are both positive at optimality, then the percentage decrease in P_i caused by spending more on component i will equal the percentage increase in P_j caused by spending less on component j.

Since the objective function for this problem is not necessarily convex, there may be multiple local minima. Thus, the analysis above does not tell us very much about the location of the global minimum. However, the key point to note is that the location of the optimum (at least for those components for which $C_i > 0$ at optimality) depends critically on the cost-effectiveness of defending the various components, as given by the derivatives of the P_i. Thus, defenders have the flexibility to allocate preventive resources to get the "most bang for the buck," and in particular the option of not defending some components at all (e.g., if they are too costly to defend).

Unconstrained case. Here, the defender can spend as much as is justified to enhance the system security. In this case, we assume that the defender wishes to minimize the expected dollar impact of an attack (in terms of the lost value of the system) plus the cost of the protective investments; i.e., $F(C_1, C_2) = \alpha\, P_1(C_1)\, P_2(C_2) + C_1 + C_2$. To find the local optima for this function, we apply partial differentiation with respect to C_1 and C_2, respectively, and equate the first derivatives to zero, yielding:

$$\frac{\partial F(C_1, C_2)}{\partial C_1} = \alpha P_1'(C_1)\, P_2(C_2) + 1 = 0$$

$$\frac{\partial F(C_1, C_2)}{\partial C_2} = \alpha P_1(C_1)\, P_2'(C_2) + 1 = 0$$

Thus, if the global optimum satisfies $C_1, C_2 > 0$, then at optimality we must have:

$$P_1'(C_1)\, P_2(C_2) = P_1(C_1)\, P_2'(C_2) = -1/\alpha \qquad (4)$$

which also implies (as in the constrained case):

$$\frac{P_1'(C_1)}{P_1(C_1)} = \frac{P_2'(C_2)}{P_2(C_2)}$$

In Equation (4), $P_1'(C_1)\, P_2(C_2)$ is the marginal effect of an incremental unit of expenditure on component 1 in reducing the probability of success of an attack on the network. Thus, at any local optimum satisfying $C_1, C_2 > 0$, the marginal effect of an incremental unit of expenditure is inversely proportional to the value of the system.

To determine whether a critical point satisfying Equation (4) is a minimum, we apply second-order differentiation to the objective function, yielding:

$$\frac{\partial^2 F(C_1, C_2)}{\partial C_1^2} = \alpha\, P_1''(C_1)\, P_2(C_2)$$

$$\frac{\partial^2 F(C_1, C_2)}{\partial C_2^2} = \alpha\, P_1(C_1)\, P_2''(C_2)$$

$$\frac{\partial^2 F(C_1, C_2)}{\partial C_1 \partial C_2} = \alpha\, P_1'(C_1)\, P_2'(C_2)$$

Therefore, the Hessian matrix for this objective function is:

$$H = \begin{vmatrix} \alpha P_1''(C_1)\, P_2(C_2) & \alpha\, P_1'(C_1)\, P_2'(C_2) \\ \alpha\, P_1'(C_1) P_2'(C_2) & \alpha\, P_2''(C_2)\, P_1(C_1) \end{vmatrix}$$

The determinant is given by $J = \alpha^2 P_1''(C_1) P_2''(C_2) P_1(C_1) P_2(C_2) - \alpha^2 [P_1'(C_1) P_2'(C_2)]^2$. Substituting in the values of $P_1'(C_1)$ and $P_2'(C_2)$ from Equation (4), we get:
$$J = \alpha^2 P_1''(C_1) P_2''(C_2) P_1(C_1) P_2(C_2) - \alpha^2 \{1/[\alpha^2 P_1(C_1) P_2(C_2)]\}^2$$
$$= \alpha^2 P_1''(C_1) P_2''(C_2) P_1(C_1) P_2(C_2) - 1/[\alpha P_1(C_1) P_2(C_2)]^2$$

A critical point satisfying the first-order conditions will be a local minimum when J is positive; i.e., when $P_1''(C_1) P_2''(C_2) > 1/[\alpha^4 P_1(C_1)^3 P_2(C_2)^3]$, or in other words, when the parameter α is sufficiently large. When α is small enough, then no critical point will be a local minimum, and the global minimum will occur with C_1 and/or C_2 equal to zero. As before, this can also occur when there are multiple critical points.

Special cases.
1. Let $P_i(C_i) = a_i e^{-b C_i}$. Here, the first-order condition for optimality holds for all values of C_1 and C_2 satisfying $C_1 + C_2 = \ln[-1/(\alpha a_1 a_2 b)]/b$, and any point on this line segment will be a global minimum. Thus, as α grows, the optimal level of defensive investment will grow, which is intuitively reasonable.

2. Let $P_i(C_i)$ be log-convex. As in the constrained case, the objective function will now be convex, ensuring the existence of a unique local minimum (possibly with C_1 and/or C_2 equal to zero, depending on the value of α).

3. Assume now that the functions $P_i(C_i)$ are identical, so that the objective function can be written as $F(C_1, C_2) = \alpha P_1(C_1) P_1(C_2) + C_1 + C_2$. In this case, if the global optimum has both C_1 and C_2 positive, then by the results derived above, they must satisfy:

$$\frac{P_1'(C_1)}{P_1(C_1)} = \frac{P_1'(C_2)}{P_1(C_2)}$$

Thus, there will exist a critical point with $C_1 = C_2$. However, this will not necessarily be the global minimum unless the objective function is convex.

Extension to systems with more than two components. In this case, the success probability of an attack is $P(C_1, C_2, ..., C_n) = P_1(C_1) P_2(C_2) ... P_n(C_n)$; the objective function is $F(C_1, C_2....C_n) = \alpha [P_1(C_1) P_2(C_2) ... P_n(C_n)] + C_1 + C_2 + ... + C_n$. To find the optimum, we differentiate the objective function with respect to C_i and equate the result to zero, yielding:

$$\alpha P_i'(C_i) \prod_{j \neq i} P_j(C_j) + 1 = 0 \qquad (5)$$

or in other words:

$$P_i'(C_i) = -1/[\alpha \prod_{j \neq i} P_j(C_j)] \qquad (6)$$

As before, there can be multiple critical points, and/or a global minimum at which one or more of the C_i are equal to zero and hence need not satisfy Equations (5) and (6). However, if α is sufficiently large that both C_i and C_j are non-zero at the global minimum, then at optimality the percentage rates of change of the success probabilities of attacks against those components will be equal, and we will have:

$$\frac{P_i'(C_i)}{P_i(C_i)} = \frac{P_j'(C_j)}{P_j(C_j)} = -1/[\alpha \prod_{i=1}^{n} P_j(C_j)]$$

In other words, the percentage rates of change in the success probabilities of attacks against components i and j will be equal, and will be inversely proportional to the expected losses due to attacks on the system.

As the value of α decreases, Equation (5) will hold for fewer components at optimality, and the optimal solution will be to invest no resources at all in the defense of some components. When α is sufficiently small, the objective function will be monotonically increasing for all non-negative values of the C_i, and at optimality no resources will be devoted to system defense.

Interpretation of results. For sufficiently small values of α, it will not be cost-effective to invest resources to defend any components, and the global minimum will be $C_i = 0$ for all i. As α becomes larger, it will eventually become worthwhile to invest in the defense of one or more components. As in the constrained case, since the objective function is not necessarily convex, there can be multiple local minima.

At any local optimum satisfying $C_i > 0$, the marginal effect of additional investment in component i will have a constant value of $-1/\alpha$ (although this need not be true for components that are not defended at optimality). Therefore, as in the constrained case, the optimal allocation of resources for defense depends critically on the cost-effectiveness of defending the various components, as given by the derivatives of the P_i. As in the constrained case, the optimal solution may involve not defending one or more components if they cannot be defended cost-effectively (relative to the value of the system). Unlike in the constrained case, the optimal solution may also involve not defending any components if α is sufficiently small.

Components in Series

Here, we consider a general system with n components in series. In a series system, if even one of the components is disabled, the entire system fails. This can occur for several reasons. In some cases, components may be physically in series (as might be the case, for example, for long pipelines or electricity transmission lines). However, one can also view multiple failure modes for a single component or system as being conceptually in series.

We again assume that the defender wants to achieve either the optimal security within a budget constraint, or the optimal tradeoff between security and the cost of defenses, taking into account the optimal attacker behavior. However, since the components are now in series, the attacker no longer has to disable all components in order to succeed, but instead has a choice of which component(s) to attack. We assume that the attacker chooses the attack strategy that will maximize the probability of the attack successfully disabling the system. The optimal defense strategy will depend on the attacker's constraints and level of knowledge about the system. Thus, we consider a set of nested optimization problems with varying combinations of assumptions about the attacker and the defender, as described below.

We assume that the attacker is limited to a single attack, and can target only a single component in that attempt. While clearly not fully general, this can be a realistic model in some circumstances (e.g., because of attacker resource constraints, or if a single failed attack would lead to the attacker being detected and disabled). In addition, multiple attacks against a single component can be modeled as a "compound" attack, with a correspondingly greater probability of success. We consider sub-cases in which the attacker has either perfect knowledge of the system's defenses or no knowledge, and in which the defender has a budget constraint or not.

Perfect attacker knowledge of defenses. We assume first that the attacker has perfect knowledge of the system's defenses. This can be a realistic assumption in some cases; e.g., if the attacker has access to inside information. In this case, the optimal attack strategy is to target the weakest component. Thus, the security of the system as a whole is equal to the security of the weakest component. Mathematically, this can be represented by $P(C_1, C_2, ..., C_n) = \max[P_i(C_i)]$. Here, the defender's resources are best utilized if the $P_i(C_i)$ are brought equal to each other (or as close to equal as possible while satisfying the budget constraint). To see this, consider first the case of two components in series.

Suppose the initial probabilities of attacks succeeding against components 1 and 2 before any defensive resources have been spent satisfy $P_1(0) < P_2(0)$. In this case, since the attacker is assumed to have perfect information about the security of the system, the total probability of a successful attack on the system will be $\max[P_i(0)] = P_2(0)$. Thus, any investment that the defender makes to reduce the success probability of an attack on component 1 will not make any difference to the total security of the system, and hence will not be optimal.

From the above, it is clear that in this case, defensive investments should be allocated so that the success probabilities of attacks against the various components $P_i(C_i)$ are equal to each other (or as close as possible within the budget constraint). Thus, resources should initially be used solely to bring the success probability of an attack against the weaker component to par (or as close to par as possible) with the success probability of an attack against the stronger component—i.e., the component with the lowest initial success probability $P_i(0)$. The extent to which this can (or should) be done will depend on the resource constraint of the defender (or the dollar value α of improving security), since it may be infeasible (or non-optimal) to achieve equality of the success probabilities of attacks against the various components.

To obtain a better understanding about the optimal allocation of defensive investments in this case, we now give a detailed mathematical formulation for optimal defense of a series system against an informed adversary. We consider both constrained and unconstrained defender optimization problems.

Perfect attacker knowledge—constrained case. In this case, the defender has a finite budget C available for strengthening the system, and the objective function to be minimized is $F(C_1, C_2) = \max[P_1(C_1), P_2(C_2)]$. Note that the objective function is monotonically decreasing. Therefore, the minimum will occur where $C_1 + C_2 = C$.

We assume without loss of generality that initially, the probability of an attack against component 1 succeeding is less than that of an attack against component 2; i.e., $P_1(0) < P_2(0)$. Define C_2^* such that $P_1(0) = P_2(C_2^*)$; i.e., $C_2^* = P_2^{-1}[P_1(0)]$. Thus, C_2^* is the level of expenditure required to equalize the success probabilities of attacks against the two components.

If the budget constraint is such that C $\leq C_2^*$, then the optimal strategy for the defender is to use all available resources for improving component 2. In this case, $P_2(C)$ will still be greater than $P_1(0)$ when the budget constraint is reached, so the defender should not strengthen component 1 at all.

If budget constraint C exceeds C_2^*, the defender should first spend an amount C_2^* to strengthen component 2 until it is at par with component 1, and then invest further in both components in such a manner that the success probabilities of attacks against the two components remain at par with each other. At optimality, we will have $P_1(C_1) = P_2(C_2)$. Thus, by the monotonic nature of the objective function, the optimum will occur at the point satisfying $P_1(C_1) = P_2(C - C_1)$, where $C - C_1 > C_2^*$.

Extension to systems with more than two components. As noted above, an attacker with perfect information about system defenses must target the weakest component in order to maximize the probability of an attack succeeding. The defender therefore must again allocate the defensive resources in such a manner that the $P_i(C_i)$ are equal to each other (or as close to each other as possible given the budget constraint).

Because the objective function is monotonically decreasing, we know that the constrained minimum will occur at $C_1 + C_2 + \ldots + C_n = C$. The resources of the defender must initially be utilized to try to bring the weakest component—i.e., the component with the highest initial success probability of an attack, $P_i(0)$—to par with the second weakest component. Then, if that can be achieved, those two components can be strengthened to bring them to par with the third weakest (or as close as possible), and so on until the resource constraint is reached. If the defender is able to bring all components to par with the strongest component, any remaining resources should then be allocated in such a manner that the $P_i(C_i)$ remain equal.

Interpretation of results. The above discussion indicates that if the attacker knows about the system's defenses, the defender's options for protecting a series system are limited. In particular, the attacker's ability to respond strategically to the defender's investments deprives the defender of the ability to allocate defensive investments according to their cost-effectiveness (as measured, for example, by the slopes of the P_i). Thus, when components are in series, and potential attackers know (or can easily learn) about the effectiveness of any defensive measures, defensive investments must essentially equalize the strength of all components in order to be beneficial.

Dresher [1961] earlier noted in the military context that, for optimal allocation of defensive resources, "It is necessary that each of the defended targets yield the same payoff to the attacker." Thus, even if some components can be hardened inexpensively, focusing protective investments on only those will lead to wasted resources if adversaries can choose to attack components that are more costly to harden. This suggests that defense will generally be more costly when the adversary knows about the system defenses. Similarly, in an analysis of asymmetric warfare against civilian targets, Ravid [2001] argues that when the adversary can change targets in response to defensive investments, "investment in defensive measures, unlike investment in safety measures, saves a lower number of lives (or other sort of damages) than the apparent direct contribution of those measures." He concludes that security improvements will tend to be more costly than safety improvements.

Perfect attacker knowledge—unconstrained case. Here, the defender has no budget constraint. As before, the constant α expresses the dollar value of reducing the success probability of an attack against the system from one to zero, and we assume that the defender wishes to optimize the tradeoff between security and cost allocation. In this case, the probability that an attack against the system will succeed remains $P(C_1, C_2) = \max[P_1(C_1), P_2(C_1)]$. However, the objective function now takes into account the tradeoff α and the costs C_i:

$$F(C_1, C_2) = \alpha \max [P_1(C_1), P_2(C_2)] + C_1 + C_2 \qquad (7)$$

This function is convex. Thus, any local minimum must be a global minimum. However, the optimal solution can take on several different forms. For sufficiently small values of α, it will not be cost-effective to invest resources to defend either of the components, and the global minimum will occur at $C_1 = C_2 = 0$. As α becomes larger, it will eventually become worthwhile to invest in defense.

If the global minimum has both $C_i > 0$, then we must have $P_1(C_1) = P_2(C_2)$; i.e., $C_2 = P_2^{-1}[P_1(C_1)]$. Substituting this expression into Equation (7), we can now rewrite the objective function in the form $F\{C_1, P_2^{-1}[P_1(C_1)]\} = \alpha P_1(C_1) + C_1 + P_2^{-1}[P_1(C_1)]$. While the original objective function was not differentiable, the revised objective function is differentiable for all non-negative C_1 such that $P_2^{-1}[P_1(C_1)]$ is greater than zero. Moreover, since the objective function is convex, differentiating with respect to C_1 and setting the result equal to zero yields a global minimum satisfying:

$$\frac{F\{C_1, P_2^{-1}[P_1(C_1)]\}}{\partial C_1} = \alpha \frac{\partial P_1(C_1)}{\partial C_1} + 1 + \frac{\partial P_2^{-1}[P_1(C_1)]}{\partial P_1(C_1)} \frac{\partial P_1(C_1)}{\partial C_1}$$

$$= \alpha P_1'(C_1) + 1 + P_2^{-1'}[P_1(C_1)] P_1'(C_1) = 0$$

Therefore, if both of the C_i are positive at optimality, then the optimal solution will satisfy $P_1'(C_1) = -1/\{\alpha + P_2^{-1'}[P_1(C_1)] P_1'(C_1)\}$.

Special case. Assume that the $P_i(C_i)$ are identical; i.e., $P_1(C_1) = P_2(C_1)$. Thus, we can now write the objective function as $F(C_1, C_2) = \alpha \max[P_1(C_1), P_1(C_2)] + C_1 + C_2$. If α is sufficiently large to justify some non-zero level of defensive investment, then the optimum will occur when $P_1(C_1) = P_1(C_2)$; i.e., when $C_1 = C_2$. In this case, rewriting the objective function in terms of C_1 alone yields $F\{C_1, P_1^{-1}[P_1(C_1)]\} = \alpha P_1(C_1) + 2$

C_1. Differentiating and setting the result to zero, we find that the optimum will occur at the point satisfying $P_i'(C_i) = -2/\alpha$.

Extension to systems with more than two components. We have assumed that the attacker possesses perfect information regarding the system defenses, and will always target the weakest component. Therefore, regardless of the number of components, the defender should allocate the defensive resources in such a manner that the $P_i(C_i)$ are brought equal to each other, or as near to equal as can be justified given the value of α. As in the two-component case, the cost-effectiveness of strengthening the various components will affect the total of the defensive investments, but not the allocation of that total investment among the various components.

Interpretation of results. Unlike in the parallel case, the optimal defensive strategy does not depend strongly on the cost-effectiveness of strengthening the various components. While the optimal solution does depend on the derivatives P_i' (and also the derivatives of the P_i^{-1}), those derivatives now influence only the total level of investment $C_1 + C_2$, since beyond some point, further increases in defensive investments will no longer be justified. Thus, the magnitude of the total defensive investment depends on the value of α, with small values of α leading to small or even zero total investment. However, the goal of bringing the weaker components to par with the stronger ones will completely dictate the allocation of any non-zero investment among the various components.

No attacker knowledge of defenses. When the attacker has no information about the system's defenses, we assume that the attacker will target one component at random, without regard to the defensive investments C_i. Consider a system with two components, and suppose that any attack will target component 1 with probability q and component 2 with probability $1 - q$. The success probability of an attack will be $P(C_1, C_2) = q\, P_1(C_1) + (1 - q)\, P_2(C_2)$. $P(C_1, C_2)$ is clearly convex, since $P_1(C_1)$ and $P_2(C_2)$ are convex, and q and $1 - q$ are non-negative. As before, we consider both constrained and unconstrained cases.

No attacker knowledge—constrained case. Here, the defender is assumed to have a resource constraint C, and the objective function to be minimized by the defender is $F(C_1, C_2) = P(C_1, C_2) = q\, P_1(C_1) + (1 - q)\, P_2(C_2)$. This function is monotonically decreasing, so the optimum will occur at $C_1 + C_2 = C$. To find the minimum, we substitute $C_2 = C - C_1$ into the objective function, differentiate with respect to C_1, and equate the result to zero, yielding:

$$\frac{\partial F(C_1, C - C_1)}{\partial C_1} = q\, P_1'(C_1) - (1 - q)\, P_2'(C - C_1) = 0$$

By convexity, any local minimum must be the unique global minimum. Moreover, if the local minimum occurs for C_1 between zero and C, it will satisfy $q\, P_1'(C_1) = (1 - q)\, P_2'(C - C_1)$. The optimum can also occur with C_1 equal to zero if component 2 can be defended much more cost-effectively than component 1, or with C_1 equal to C if the converse, in which case it need not satisfy this first-order condition.

Special case. Let the conditional probability that each component is attacked be 0.5. To optimize the system defenses, the defender should achieve $P_1'(C_1) = P_2'(C - C_1)$ if this can be achieved for $0 < C_1 < C$. At this point, the improvement in the security of component 1 by increasing C_1 will be exactly equal to the corresponding deterioration in the security of component 2. Thus, when the conditional probabilities of attacks on the two components are equal, the defender should invest more in the component that is more cost-effective to strengthen. Moreover, if the functions P_i are identical, then the defender should spend the same level of resources on each component.

Extension to systems with more than two components. For a system with n components, the objective function that the defender wishes to minimize is:

$$F(C_1, C_2, \ldots, C_n) = \sum_{i=1}^{n} q_i P_i(C_i)$$

where q_i is the probability that an attack will target component i, and the q_i are assumed to sum to one. The global minimum will occur at $C_1 + C_2 + \ldots + C_n = C$, or in other words $C_1 = C - (C_2 + \ldots + C_n)$. Substituting this value of C_1 into the objective function, we get:

$$F(C_1, C_2, \ldots, C_n) = q_1 P_1[C - (C_2 + \ldots + C_n)] + \sum_{i=2}^{n} q_i P_i(C_i)$$

To find the global optimum, we differentiate the objective function with respect to C_i (for $i \neq 1$) and equate the result to zero, yielding:

$$\frac{\partial F(C_1, C_2, \ldots, C_n)}{\partial C_i} = q_1 P_1'[C - (C_2 + \ldots + C_n)] + q_i P_i'(C_i) = 0$$

At optimality, we will have $q_1 P_1'[C - (C_2 + \ldots + C_n)] = q_i P_i'(C_i)$, if this occurs for positive values of the C_i. Thus, the marginal effects of additional resources spent on all defended components will be equal; i.e., components with C_i non-zero will all have the same marginal effectiveness for additional investment. This is similar to the result for a two-component system. The level of resources spent on component i will also generally increase with the conditional probability of attack q_i.

Interpretation of results. The above results show that the defender has much more flexibility to allocate resources between the components of a series system cost-effectively when the attacker has no knowledge of the system defenses than when the attacker has perfect information about the defenses. In other words, for a fixed budget, the defender will generally be able to achieve better system security when the attacker has little or no knowledge of the system defenses. This demonstrates the potential importance of secrecy (and/or deception) as defensive strategies.

As one example, the proposal to sterilize mail to protect against future anthrax attacks [Florig 2002] might be an effective defense if the installation of such sterilization equipment were a secret. However, this would probably not be possible in our open society, given the outcry of public concern about the threat of anthrax spores sent through the mail. If the installation of anthrax sterilization equipment were public knowledge, it would likely only cause future attackers to find a means of delivery other than the public mail. In that case, the proposed large investment in sterilization equipment may never sterilize a single anthrax spore!

No attacker knowledge—unconstrained case. Here, the defender has no budget constraint, but the parameter α gives the importance assigned to system security relative to the cost of defensive investments. The probability of an attack succeeding remains $P(C_1, C_2) = q\, P_1(C_1) + (1 - q)\, P_2(C_2)$. However, the objective function must now take into account the tradeoff α and the costs C_i, yielding $F(C_1, C_2) = \alpha\, [q\, P_1(C_1) + (1 - q)\, P_2(C_2)] + C_1 + C_2$. The objective function $F(C_1, C_2)$ is still convex, since α is non-negative and the cost terms are linear. To find the global minimum for this function, we differentiate with respect to C_1 and C_2, respectively, and equate the first derivatives to zero:

$$\frac{\partial F(C_1, C_2)}{\partial C_1} = \alpha\, q\, P_1'(C_1) + 1 = 0$$

$$\frac{\partial F(C_1, C_2)}{\partial C_2} = \alpha\, (1 - q)\, P_2'(C_2) + 1 = 0$$

By the convexity of the objective function, if the global minimum occurs where C_1—respectively, C_2—is positive, then at optimality, we will have $P_1'(C_1) = -1/(\alpha\, q)$—respectively, $P_2'(C_2) = -1/[\alpha\, (1 - q)]$. However, for sufficiently small values of α, the global minimum may have one or both of the C_i equal to zero, in which case the above condition need not be satisfied for the variable(s) equal to zero at optimality.

Special cases.
1. As in the constrained case above, if the conditional probability that each component is attacked equals 0.5, then the defender should invest more resources in the component that is more cost-effective to strengthen.

2. When the functions $P_i(C_i)$ are identical, the defender should invest money in the component that the attacker is more likely to target, as is intuitive.

Extension to systems with more than two components. As above, for any component i receiving a non-zero defensive investment at optimality, we will have $P_i'(C_i) = -1/(\alpha\, q_i)$, where q_i is the conditional probability of an attack on component i. Thus, the results are similar to those for two components.

Interpretation of results. Increases in the value of α will lead to increases in the optimum values of the C_1, at least for those components that are already receiving positive levels of investment at optimality. In addition, an increase in the conditional probability of an attack on a particular component will generally increase the optimum investment in that component. Also, in this case, the optimal levels of investment in the various components can be determined independently of each other, since there is no longer a budget constraint to induce coupling between them.

Conclusions and Future Work

Our results suggest that defending series systems against informed attackers is an extremely difficult challenge. This emphasizes the importance of redundancy as a defensive strategy—especially if attacks against the redundant components are also likely to succeed or fail approximately independently of each other (e.g., due to factors such as spatial separation and functional diversity). Our results also support the idea that secrecy or even deception can be an important strategy for improving security (especially for series systems), and/or reducing defensive costs.

Obviously, it is important to extend this work to more general systems, rather than only series and parallel architectures. Unfortunately, finding optimal attack strategies for arbitrary systems is NP-hard, as can be shown from results in Cox et al. [1989, 1996]. Therefore, we plan to adapt near-optimal heuristics for least-cost diagnosis from Cox et al. to derive near-optimal heuristic attack strategies, and identify optimal or near-optimal defenses against such attacks. We also hope to determine when those heuristic attack strategies will be optimal.

Considering more general system structures (e.g., using graph theory methods) would allow us to model attacks for instrumental purposes, such as internet "worms," in which poorly defended computers are compromised in order to use them in an attack against a more valuable but better defended system. One could also use similar approaches to investigate quantitatively the relative merits and limitations of perimeter defense in computer networks; see, for example, Wulf and Jones [2002].

In addition, we plan to extend our models to consider systems in which the various components have different inherent "values" to the defender (and hence to the attacker), rather than being valuable only insofar as they contribute to the operability of the system as a whole. For example, a defender might be more concerned about an attacker successfully accessing a database of critical information than about an attack that merely renders a computer system unusable. Such models would help us investigate the appropriateness of the heuristic suggested by O'Hanlon et al. [2002], of allocating resources to protect only the most valuable assets. Preliminary results suggest that the suitability of this heuristic will depend to some extent on the assumptions made about attacker goals and motivations. For example, the proposed heuristic may be optimal if attackers choose targets solely according to their value, but not if they also consider the probability of success in their choice of targets.

Also, it would be worthwhile to extend our models to include a time dimension, rather than the current static or "snapshot" view of system security. This would allow us to model imperfect attacker information (including Bayesian updating of the probability that an attack will be successful, as estimated by both attacker and defender), and the possibility of multiple attacks (either attacks against multiple components, or multiple attacks against the same component).

One could also apply similar methods to model attacks involving opportunistic rather than determined adversaries. In this case, it may be helpful to formulate the problem as a game between multiple defenders [Kunreuther and Heal 2002], rather than a game between a single defender and a single adversary.

Acknowledgments

This material is based upon work supported in part by the US Army Research Laboratory and the US Army Research Office under grant number DAAD19-01-1-0502. Any opinions, findings, and conclusions or recommendations expressed in this document are those of the authors and may reflect the views of the sponsors.

References

Cox, L.A., Y. Qiu, and W. Kuehner (1989). "Heuristic least-cost computation of discrete classification functions with uncertain argument values." *Ann. of Operations Research*, 21, 1-30.

Cox, L.A., S. Chiu, and X. Sun (1996). "Least-cost failure diagnosis in uncertain reliability systems." *Reliability Eng. and System Safety*, 54(2-3), 203-216.

Dresher, M. (1961). *Games of Strategy: Theory and Applications*. Englewood Cliffs, NJ: Prentice-Hall.

Florig, H.K. (2002). "Is safe mail worth the price?" *Science*, Feb. 22, 1467-1468.

Kunreuther, H. and G. Heal (2002). "Interdependent security: the case of identical agents." Insurance Project Workshop, National Bureau of Economic Research, Inc., Cambridge, MA, http://www.nber.org/~confer/2002/insw02/kunreuther.pdf.

O'Hanlon, M., P. Orszag, I. Daalder, M. Destler, D. Gunter, R. Litan, and J. Steinberg (2002). *Protecting the American Homeland*. Washington, DC: Brookings Institution.

President's Commission on Critical Infrastructure Protection (1997). *Critical Foundations: Protecting America's Infrastructures*. Washington, DC, http://www.terrorism.com/homeland/pccipreport.pdf.

Ravid, I. (2001). "Theater ballistic missiles and asymmetric war." Draft from author.

Schneier, B. (2001). "Military history and network security" (sidebar). Counterpane Internet Security, Cupertino, CA, http://www.counterpane.com/msm.pdf.

Schudel, G. and B. Wood (2000). "Modeling behavior of the cyber-terrorist." In Anderson, R.H., T. Bozek, T. Longstaff, W. Meitzler, M. Skroch, and K. Van Wyk, *Conference Proceedings: Research on Mitigating the Insider Threat to Information Systems—#2*. Santa Monica, CA: Rand.

Wulf, W.A. and A.K. Jones (2002). "Cybersecurity." *The Bridge*, 32(1), 41-45.

Applying the General Theory of Quantitative Risk Assessment (QRA) to Terrorism Risk

Stan Kaplan[1]

Abstract

The purpose of this paper is to point out that the general theory of quantitative risk assessment (QRA) applies perfectly well to evaluating and quantifying the risk from terrorism. The main difference occurs during the "scenario identification" part of the risk assessment process. Whereas, in an "ordinary" QRA, we ask the question, "What can go wrong?," in terrorism risk assessment (TQRA) we ask, "If I wanted to, what could I make go wrong?" In answering this new question, the Theory of Scenario Structuring and the use of fault and event trees play the major roles as before. Also, the concept of "resources" now moves to center stage as part of the process of identifying terrorism scenarios. So also does the use of Bayes' theorem, not only to assess *a priori* the likelihoods of specific terrorism scenarios, but also as a crucial part of surveillance systems that have, potentially, the ability to quantify the likelihoods that such scenarios are in process.

The Quantitative Definition of Risk

We begin by recalling the definition of risk, R, as a set of triplet questions [see Kaplan and Garrick 1981; Kaplan 1991]:

$$R = \{<S_i, L_i, X_i>\}_c$$

Here S_i denotes a risk scenario, L_i denotes the likelihood of that scenario, and X_i denotes the consequences of that scenario. The angle brackets thus enclose a "risk triplet" and the curly brackets are standard mathspeak for "set of." The subscript c denotes a suitable space over which the subscript index, i, varies. For practical purposes c is taken to be a finite set. The task of ordinary QRA is thus to identify the S_i and for each one, to quantify the likelihood L_i and consequence X_i.

---

[1]Center for Risk Management of Engineering Systems, University of Virginia, Charlottesville, VA 22903; 310-377-2519; stankap@aol.com.

The "As-Planned" Scenario S_0

Typically (but not necessarily), the functioning of an "as-planned" system can be represented by a linear chain of steps or "operations" as shown in Figure 1. These steps culminate in the "success state" ES_0.

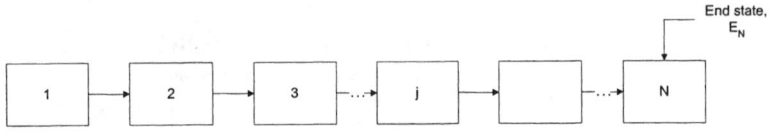

Risk analysis question: What can go wrong here?

Terrorism question: What can I make go wrong here?

Figure 1. The "as-planned" or "success" scenario, S_0.

Ordinary Risk Analysis

In ordinary risk analysis, therefore, we look at each step and ask, "What can go wrong here?" That something that goes wrong is termed an *initiating event* (IE) in Figure 2. Emanating from that IE is an *event tree* (ET), representing all the possible consequences of that IE. Those consequences are labeled in Figure 2 as the *end states* ES_1, ES_2, etc. They reflect the damage resulting from the IE. Therefore, in ordinary risk analysis we ask for each stage, "What can go wrong here?," "How likely is that to happen?," and, "If it does happen, what are the consequent end states?"

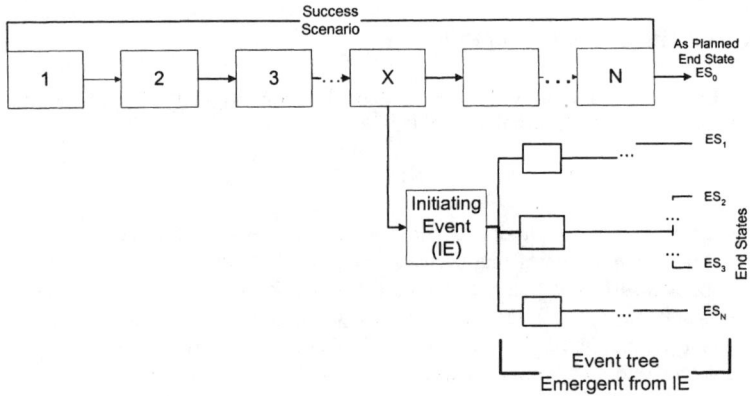

Figure 1. Event tree emanating from an initiating event (IE).

Terrorism Risk Analysis: Identifying Possible Risk Scenarios

In terrorism risk analysis, the Theory of Scenario Structuring [Kaplan et al. 1999] and the use of fault and event trees take center stage. As in "ordinary" risk analysis, the process begins by clearly identifying the "target system" and describing for that system the "as-planned," or "success," scenario, S_0, as in Figure 1. If all the stages succeed, then the process succeeds, by definition. For the process to fail then, one or more of the stages must fail. Therefore, we look at each stage as terrorists would and ask ourselves, "How can I create a failure of this system in this stage?" [Kaplan et al. 1999].

The answers to that question may be envisioned as a fault tree for that stage, as shown in Figure 3. Such a tree, if done properly, may be considered to portray all the possible ways that the stage in question can be made to fail. It also should portray the resources that terrorists must have and apply in order to make that stage (or stages) fail. "Resources" is used in a very general sense here, and could include explosives, poisons, skills, people, information, access, and other factors.

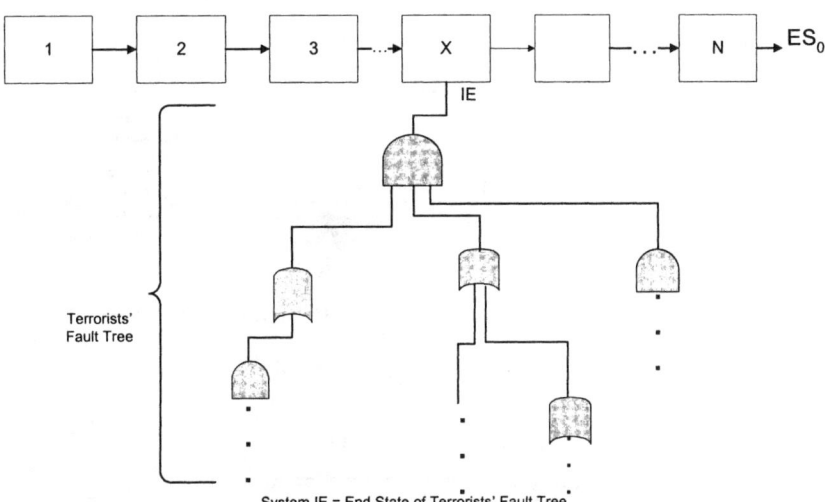

Figure 2. Terrorist fault tree culminating in initiating event.

Detecting Terrorist Attacks in Preparation

From the data represented in Figures 1, 2, and 3, we can identify the significant terrorist scenarios, i.e. terrorist attacks, and for each, calculate the likelihood of its success. We can also see what the terrorists must do and what resources they must acquire in order to carry out these attacks. That opens up the possibility of developing surveillance systems, as shown in Figure 4, which could detect signs and symptoms of a terrorist attack in preparation. Those signs and symptoms, their presence, absence, degree, timing, etc., constitute items of evidence which could be fed continuously into a Bayesian logic engine which would continuously calculate the probability that various attack scenarios are in preparation.

Conclusions and Commentary—Surveillance Systems

The linear chain structure of S_0, depicted in Figure 1, may be a simplification of the actual situation, but this is not important. If the actual structure of the "as-planned" scenario is more complicated, that fact can be easily reflected in a more realistic Figure 1 diagram, and the remainder of the analysis proceeds as before with no problem. Indeed, Figures 1 through 4 capture the essential ingredients of the problem and cannot be omitted or altered, other than cosmetically. A particularly attractive idea is a surveillance system with sensors feeding a central computer containing a Bayesian logic engine (see Figure 4). This engine could be designed to print out, periodically and upon request, probability curves showing the likelihoods of various possible attacks, and by whom, when, and where.

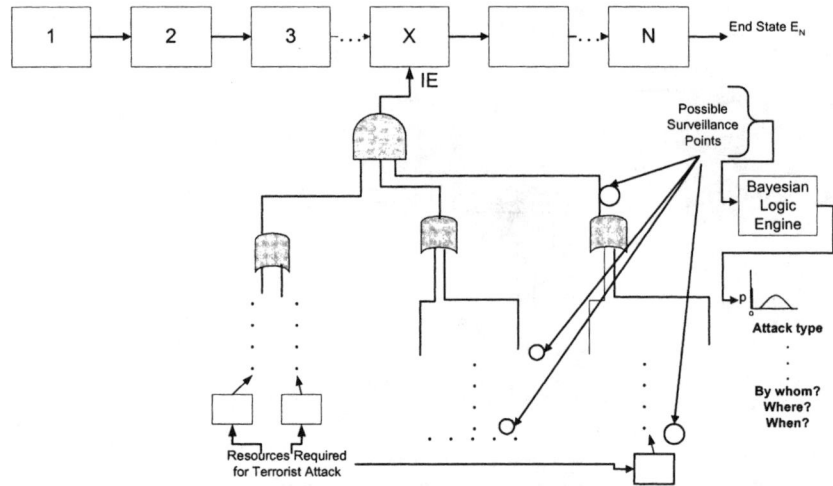

Figure 3. Use of surveillance points and Bayesian logic to detect terrorist scenarios in preparation.

If the analysis shows that our system is vulnerable to a particular type of attack, then we must take action to reduce the likelihood of such an attack, to detect it during its preparatory stages, or to reduce the damage consequences if the attack does occur. Preferably, all of the above. There is no other way to protect ourselves, our families, and our nation.

References

Kaplan, S. and B.J. Garrick (1981). "On the quantitative definition of risk." *Risk Analysis,* 1(1), 11-27.

Kaplan, S. (1991). "The general theory of quantitative risk assessment." *Risk-Based Decision Making in Water Resources V,* 11-39. American Society of Civil Engineers, Reston, VA.

Kaplan, S., B. Zlotin, A. Zussman, and S. Vishnipolski (1999). *New Tools for Failure and Risk Analysis—Anticipatory Failure Determination and the Theory of Scenario Structuring.* Monograph published by Ideation Inc., Southfield, MI.

Vulnerability of Water Systems to Acts of Terrorism and Acts of Nature

Nicholas C. Matalas[1]

Abstract

The vulnerability of the nation's water infrastructure to terrorist threats is discussed, and the differences and similarities in the vulnerabilities to acts of terror and to acts of nature are noted. Haimes et al [1998] proposed hardening water resource systems by adding or augmenting security systems and by enhancing the redundancy, robustness, and resiliency of the systems. A general form of hardness perhaps may be given to the systems through a national water board that would coordinate the hardening activities of federal agencies and their responses to crises. Decisions depend on future expectations. The decision made to harden the infrastructure will be made in a state of uncertainty, given that the future cannot be clearly comprehended. The traditional presumption that uncertainty derives from imperfect knowledge and therefore can be addressed through the theory of probability may not be sufficient in the face of terrorist threats. Some decisions might be made in a state of uncertainty that derives from incomplete knowledge; this may be addressed through a theory of surprise.

Introduction

Apart from its people, water is perhaps the most vital resource of any nation. The United States is among the water-rich countries of the world, others being Brazil, India, Canada, and Russia. However, water richness does not translate into water security. The nation's water resource systems that have always been vulnerable to acts of nature are now also vulnerable to acts of terror. Terror, in this sense, is a state of intense fear resulting from threats or violence perpetrated by one or more individuals to intimidate specific persons, segments of society, or governments into meeting the terrorists' demands.

Water resource systems, as well as other infrastructures, cannot be rendered totally invulnerable to terrorist acts or to acts of nature. The spatial scale of the collection of systems precludes the nation's ability to pay for hardening the systems to render them totally invulnerable to any and all acts. As the threat of terrorist attacks

---

[1]Hydrologist, 709 Glyndon St. SE, Vienna, VA 22180; 703-319-4016; nmatalas@aol.com

against water systems is real, protecting the systems against them must be considered alongside traditional objectives of water management that are achieved through balancing benefits and costs.

Terrorist attacks against water resource systems have the potential to seriously compromise the general welfare and national security of the nation. Because concern about terrorism is high on the national agenda, the vulnerability of a water system to an act of terror cannot be viewed in the same way as its vulnerability to an act of nature. Nature may hold little in the way of surprise, however uncertain the extremity of future occurrences such as floods and droughts. But surprise is the strong suit of terrorism. Terrorism-induced surprise not only relates to the where, when, and how of an attack, but also to the uncertain social reaction and consequences. The nature of the uncertainty that pervades the management of water resources is not necessarily the same in the absence of terrorism as it is when terrorism is a factor. The differences in the uncertainties impact assessing the vulnerabilities of water resource systems.

All efforts at countering terrorism are constrained, explicitly or otherwise, by maintaining the United States as an open society and not compromising its citizens' constitutional rights. These constraints, though they compound the problem of countering terrorism, usually are not subject to relaxation.

Sources of Vulnerability of Water Resource Systems

Traditionally, water resource systems have been managed to achieve a number of objectives for enhancing the well-being of society. Among the objectives is the protection of the public from potential ravages of floods and droughts. The decisions that are made to achieve the objectives are based on future expectations. Because the future cannot be clearly comprehended, the decisions are made in a state of uncertainty.

Among the future expectations are that the benefits will exceed the costs. Howe [1971] described benefits and costs as being of four types: 1) those for which market prices exist, reflecting social values, e.g., non-price-supported commodities; 2) those for which market prices exist, but the prices do not reflect social values, e.g., price-supported commodities such as cotton and wheat; 3) those for which no market prices exist, but for which social values in monetary terms can be measured in terms of willingness to pay, e.g., water-based recreation; and 4) those for which it would be difficult to imagine any kind of market-like process capable of registering meaningful monetary valuation, e.g., maintenance of a beautiful view.

How the benefits and costs entailed in countering the threat of terrorist acts are perceived and evaluated depends on a number of factors. One factor, the nature of system vulnerability, is addressed herein. In particular, the distinctions between vulnerability to acts of terror and acts of nature are noted.

Following are several sources of vulnerability:

Water resource systems are readily accessible by the public. There are tens of thousands of miles of river courses and thousands of miles of lake and wetland shorelines in the United States. These water bodies are readily accessible at almost any point along their courses or shorelines. Natural barriers, such as swamps, rugged terrain, and dense vegetation, and man-made barriers, such as fencing and posting "no trespassing" signs, may inhibit someone from accessing a water body, but they do not restrict access *per se*. A system is open to almost any sort of terrorist act. This source of vulnerability is inconsequential with regard to acts of nature.

Regional water resource systems are not integrated. One region's water system cannot be depended upon to respond effectively to another region's water crisis, however close the regions may be. Even if two regional systems are hydraulically connected, they are not likely to be connected managerially because of the lack of structural connections, and the lack of legal and institutional authority for moving water from one region to another. The water resources of a region are seldom developed with the objective of being able to respond to a short-term water crisis in a nearby region. Acts of nature and of terrorism share this source of vulnerability..

Much of the nation's water-related infrastructure is aged. The scientific community and the media have noted the poor condition of many of the nation's bridges, culverts, and spillways. These bridges and culverts are along the interstate highway system, initiated in 1952, and along the older highway system, initiated *circa* 1920. A survey of these structures, along with recommendations and cost estimates for their repair or replacement would be in order even if there were no threat of terrorism. Terrorism and nature compound any weakness that age imposes on the structures.

Water management is a diffused process on a national scale. The costs and benefits of water management are not fully integrated into national input-output accounts. The estimated costs and projected benefits associated with the development of the water resources of a region are not assessed on a national scale relative to the costs and benefits of other regional and national programs. Seldom are regional water programs assessed relative to one another. Consequently, the costs of reducing the vulnerabilities of water systems to "acceptable" levels cannot be readily weighed relative to the costs of reducing the vulnerability of other infrastructures to "acceptable" levels.

Potential Terrorist Acts

Potential terrorist acts are noted in reference to four specific purposes of a water resource system: namely, the provision of public water supply, irrigation water, hydroelectric power, and navigation. The sources of system vulnerability almost assure that a terrorist act could be conducted successfully and the intended consequences realized. Such an act may or may not be violent. For example:

A hazardous material could be discharged into a reservoir or distribution network. For the act to be effective, the material must become dispersed among the water molecules. The material may be industrial waste heavily laden with arsenic, potassium, or sodium cyanide. It may be radioactive. An advantageous time for a terrorist to act would be when a regional drought has reduced the ability of a system's waters to dilute these materials.

Early detection of the material would not necessarily render the act ineffective. If detection brings about an interruption in the public's use of the water, the act is effective, though not as effective as it would be had illness or death triggered detection. Knowing that the water supply had been contaminated with potentially lethal material, users and non users of the water would be instilled with fear, even if illness or death had not occurred. And, if one water system has been intentionally contaminated with potentially lethal material, couldn't another be attacked?

A particular chemical or biological substance could be discharged into reservoirs whose waters are used for irrigation. The material or biological substances would either damage the crops in the field or become chemically incorporated into them, or remain on the surface. Washing the crops may not render them safe for consumption. A terrorist act of this kind would likely go undetected until the produce had reached the market and illness or death occurred.

Such an act could be devastating to the economy and to the public's morale. In addition to crop damages and the illnesses and deaths resulting from consumption of the crops, the soils could become contaminated with the contaminated waters infiltrating to the underlying aquifer, thus rendering the aquifer unusable for a protracted period of time.

Explosive devices of one kind or another could disrupt the generation of electrical power at a hydropower dam. Disrupting power generation could be accomplished by destroying the generators or perhaps the dam itself.

If the dam was providing hydroelectric power to a very large number of customers, e.g., several million, over a broad geographical region, the disruption in power generation could be catastrophic. Loss of electrical power in the region would result in water shortages, increased risk of fires, extreme stress on emergency services, and severe cutbacks in industrial, commercial, and financial operations and services. Water, as well as electrical power, would need to be imported into the region, and the quantities could severely strain the national electric power grid and the water supplies of nearby regions. Costs would be large for transferring water to the water-short region.

Along a reach of a navigable river, a bridge or a set of locks could be destroyed. Such a terrorist act might not be as economically devastating as those described above. However, whether or not lives were lost, the attack would provoke a high state of anxiety among the population. The destruction of the bridge would force vehicular traffic to be rerouted, either permanently or until a new bridge could be constructed. If the bridge had been part of a major transportation artery with a high volume of traffic, large economic costs would likely accrue because of the increased congestion, delays on alternate roads, and construction of a new bridge. The rerouting of interrupted river traffic would add to the congestion on the alternate roads.

One may question whether these acts are realizable. For example, it may not be feasible for a terrorist to inject sufficient quantities of chemical or biological agents into a water supply to override their dilution into harmless concentrations. However, in the absence of experience these harmful acts can be imagined and therefore the possibilities cannot be dismissed.

Hardening Water Resource Systems

To harden a system is to reduce its vulnerability. Haimes et al. [1998] proposed hardening water resource systems against natural and terrorist acts by incorporating security measures (fences, locks, cameras, etc.) and by enhancing the degree of three specific attributes possessed by the system. These are the attributes of redundancy, robustness, and resilience, referred to collectively as the 3Rs. The 3Rs were initially discussed by Matalas and Fiering [1977] as offering a buffer to the threat to water resource systems posed by potential changes in climate. Haimes et al. [1998] placed these attributes into the context of buffering systems to reduce their vulnerability to threats of potential terrorists acts, as well as to natural hazards.

In a *redundant* system, the functions of failed components can be taken on by other components without adversely effecting the performance of the system itself. Building redundancy into a system raises its cost but reduces the risk of system failure. *Robustness* refers to a system design's degree of insensitivity to errors in the estimates of those parameters affecting the design choice. The errors may be the result of miscalculations or of statistical sampling. Statistical sampling errors derive from the stochastic inputs to the system design, particularly the hydrologic inputs. *Resilience* is the ability of an optimally designed system to be operated technically and institutionally over the short run as a near-optimally-designed system, such that the resulting economic losses are acceptable.

All systems cannot be hardened against all imaginable terrorists acts. The cost of doing so would be prohibitive. Are some systems to be left in their present state of hardness? Are larger systems, i.e., those serving large communities, to be allocated a greater share of resources for hardening than smaller systems? Would hardening be directed to all purposes of the system or to selected purposes, e.g., those of water supply, irrigation, and hydropower, but not water-based recreation and wetlands maintenance? Addressing such questions in a manner that the public would consider equitable would provide the impetus for a general hardening of the nation's water infrastructure.

One solution would be to consider establishing a national water board to coordinate and act upon the views of the full spectrum of federal water agencies. The board's responsibility would be to provide a coordinated response to terrorism, both for the "day before" (by defining the technologies, resources, and institutions for water transfers), and for the "day after" (for implementing the water transfers). The board would help set the agenda that at best would shield the nation's water resource systems from terrorist acts, and at worst would help them be fail-safe in an attack.

A national water board, the US Water Resources Council, is already in existence. The Council was established on July 22, 1965, but assumed an inactive status on October 1, 1982. If the Council were to be reactivated, its mission could be redefined in terms of protecting the nation's water resources from terrorist acts. Whether the Council or a new entity, the water board could begin by developing an inventory of the nation's water resource systems. This would include the number of systems, their locations, and their configurations, i.e., descriptions of their structural and nonstructural measures, and the sizes of the communities served. The inventory would be a basis for allocating resources for hardening the systems.

Hardening Systems in States of Uncertainty

Decisions that are made to harden systems, and thereby to render them less vulnerable to natural or terrorist acts, are based on future expectations. Because the future cannot be clearly comprehended, the expectations may not be realized. Thus, the decisions are made in a state of uncertainty. The uncertainty derives either from the imperfection of the needed knowledge or its incompleteness, in the sense that some things are unknowable and other things cannot be known ahead of time.

It is generally assumed that uncertainty derives from imperfect knowledge. For any decision to be made, all possible outcomes can be known, but which outcome actually will follow from a decision is not known. The uncertainty as to which of the known outcomes will be realized is addressed by the theory of probability. Shackle [1949] was one of the first to consider uncertainty as deriving from incomplete knowledge. If knowledge is incomplete, then all possible outcomes are not knowable at the time a decision is made. Although numbers that would sum to unity can be assigned to the known outcomes, they cannot be interpreted as probabilities. Although the concept of uncertainty deriving from incomplete knowledge is not commonplace in making economic decisions, it should not be dismissed without consideration in view of the terrorist threats to the nation's infrastructures and the public welfare..

Shackle proposed viewing the outcomes of decisions in terms of surprise—the psychological state of mind following the occurrence of counter-expected or unexpected events. *Counter-expected events* are events that have been imagined but consciously rejected as not possible. *Unexpected events* are events that have never been imagined and thus never were assessed as possible or impossible (see Shackle [1979]). Counter-expected events occur in a state of imperfect knowledge, whereas unexpected events may occur in a state of incomplete knowledge.

Fiering and Kindler [1987] suggested, without any reference to terrorism, that the notion of surprise might provide a useful characteristic of acceptable design of a water resource system. To explore the usefulness of this notion in water resource management, they offered, as a starting point, a taxonomy of surprises that they viewed as resulting from unexpected events. They took the following taxonomy to be neither exclusive nor exhaustive:

    i) structural surprise that derives from the structural collapse of a system component;
    ii) embedded system surprise that derives from embedded errors;
    iii) hydrologic surprise that derives from changes in the catchment;
    iv) institutional surprise that derives from a new law or environmental standard that imposes an unanticipated qualitative shift in the operation of a system;
    v) informational surprise that derives from disruption of crucial information;

vi) mechanistic surprise that derives from lack of understanding how a system will respond to specific stresses; and

vii) demand surprise that derives from major demographic events that cause a precipitous shift outside the conventional range of imprecision associated with such a system.

Fiering and Kindler suggested that minimizing the likelihood of surprise might be a viable alternative to optimality as a characteristic of acceptable design of a water resource system. One might question how a measure of likelihood in its probability sense could be attached to surprise derived from an unexpected event. Nonetheless, the introduction of Shackle's ideas on surprise to water management is timely. If these ideas are viable in relation to acts of nature and the responses to them (as illustrated by Fiering and Kindler), then Shackle's notion would seem to be even more viable in relation to terrorist attacks and responses.

Unexpected events can and do occur, and the future cannot mirror the past. For example, the consequences of the 1889 Johnstown flood suggest that further consideration should be given to Shackle's notion of surprise in hardening water resource systems against terrorist acts.. Community surprise, for lack of a better term, should be added to the Fiering-Kindler list. It seems fair to say that the members of the South Fork Fishing and Hunting Club might have imagined that the club's dam on the South Fork River might collapse with subsequent loss of life, but they might not have imagined the demise of Johnstown as a major center of steel production. Heavy rains brought about the collapse of the dam, already in a weakened state through neglect of maintenance by the club members. The resulting flood claimed more than 2,200 lives and the Johnstown that followed was not the same[McCullough 1987].

Summary

Over the past several decades, considerable experience has been gained in protecting the nation's water infrastructure against acts of nature. This experience forms the basis for responding to a new and different threat to the infrastructure—potential acts of terror. However, experience in dealing with natural hazards may not be sufficient to contend with terrorist threats or attacks, as the vulnerabilities of water resource systems are not strictly the same in both cases.. For example, the openness of the systems has no bearing on their vulnerability to acts of nature, but openness, or easy access, is an invitation to terrorists.

In light of terrorist threats since 9/11, hardening the water infrastructure against future attacks is of immediate concern. Haimes et al. [1998] suggested that hardening systems against acts of terror in terms of security, redundancy, robustness, and resilience also hardens them against acts of nature. These methods do not preclude other means of hardening systems. In any case, all systems cannot be effectively hardened against all possible terrorist acts, so questions arise as to priorities: Which systems are to be hardened, and to what extent? How can they be operated in order to be able to respond effectively to water crises in nearby regions? These and other question need to be answered in a manner that the public would perceive as equitable.

Perhaps a forum should be established to address these questions. Perhaps a reactivated US Water Resources Council or a new entity could serve as a national water board to coordinate federal decisionmaking activities to harden the water infrastructure and respond to water crises. Decisions to harden the infrastructure will be made in a state of uncertainty, given that the future cannot be clearly comprehended. Furthermore, in dealing with acts of nature, water management is based on the presumption that uncertainty derives from imperfect knowledge. The uncertainty is addressed through the theory of probability. Limiting the perception of uncertainty to that deriving from imperfect knowledge may not suffice in dealing with acts of terrorists. Uncertainty may also derive from incomplete knowledge. Shackle's theory of surprise is another way to address uncertainty. .

Fiering and Kindler [1987] argued that traditional water management must consider Shackle's notions of counter-expected and unexpected events. They offered an initial taxonomy of surprise deriving from unexpected events in water management. It is obvious that this can be applied to acts of terror as well as to acts of nature. Hardening systems to reduce their vulnerabilities to terrorist attacks may also reduce their vulnerability to natural hazards. Today, the driving force in reducing vulnerability is the threat of terrorism, but water systems cannot be hardened immediately. Installing security measures such as fences, locks, and cameras can be done relatively quickly. Adding redundancy to a system through adopting uniform standards and specifications of pumps and other equipment, as suggested by Haimes et al. [1998], takes time in order to reach agreements with manufacturers. Other measures of hardening that would require large capital expenditures would take considerable time to bring into operation, e.g., adding more storage and additional pipes and canals to serve as alternate means of water transport.

The costs that would be incurred in reducing the vulnerability of the nation's water infrastructure will need to be weighed against the societal costs (medical, economic, and psychological, among others) that would be incurred as a consequence of a terrorist act. How the societal costs are to be measured and weighed against the monetary costs of hardening is an open question.

References

Fiering, M. and J. Kindler (1987). "Surprise in water-resource design." *International Journal of Water Resource Development*, 2(4), 1-10.

Haimes, Y.Y., N.C. Matalas, J.H. Lambert, B.A. Jackson, and J.F.R. Fellows (1998). "Reducing vulnerability of water supply systems to attack." *Journal of Infrastructure Systems*, 1-14 .

Howe, C.W. (1971). *Benefit-Cost Analysis for Water System Planning.* Water Resource Monograph 2, American Geophysical Union, Washington, DC.

Matalas, N.C. and M.B. Fiering (1977). *Water Resource System Planning, in Climate, Climate Change and Water Supply.* Studies in Geophysics, National Research Council, National Academy of Sciences, Washington, DC, 11-109.

McCullough, D. (1987). *The Johnstown Flood,* Second Touchstone Edition. New York: Simon and Schuster, Inc.

Shackle, G.L.S. (1949). *Expectations in Economics.* London: Cambridge University Press.

Shackle, G.L.S. (1979). *Imagination and the Nature of Choice.* Cambridge, UK: Cambridge University Press.

Toward a Systems-Based Vulnerability Assessment Methodology for Water Supply Systems

Barry C. Ezell[1]

Abstract

The purpose of this paper is to provide an in-progress review of the development of a systems-based vulnerability assessment methodology for critical infrastructures. This methodology is a supporting framework for the Infrastructure Risk Analysis Model (IRAM), where vulnerability assessment was briefly discussed but not fully developed by the authors. This paper focuses exclusively on water supply as the complex organizational system in focus. However, further research, deployment, and refinement of the vulnerability methodology may prove the applicability of the framework for military installations and electric power, among other critical infrastructures. This paper discloses three main ideas. First, there has been very little published in the way of rigorous vulnerability assessment methodologies. In fact, there is no agreed-upon definition of vulnerability. Second, vulnerability assessment appears to be an *ad hoc* checklist of things-to-do, bereft of agreement regarding the system in focus, i.e., the boundaries and context of a water supply system. Last, the paper argues that a water supply system should be viewed as a complex organizational system. The implication of this paper is that classic risk assessment questions (What can go wrong? What is the likelihood? and What are the consequences?) should be preceded by the primal question: What is the system in focus? The system in focus must be understood before meaningful risk and vulnerability assessment is undertaken.

Introduction

Military and civilian leaders have the responsibility to protect our nation's critical infrastructures, communities, and symbols of American power from terrorists, home and abroad, as well as from natural disasters. This paper introduces a systems-based

---

[1] Major, US Army, and doctoral student, Old Dominion University, Department of Engineering Management and Systems Engineering, Norfolk, VA 23529; 757-831-3632; bcezell@aol.com.

vulnerability methodology for water supply systems. The methodology is systems-based, because it begins with an understanding of what exactly is the system in focus for the vulnerability assessment. The paper points out that the risk and infrastructure systems literature review reveals a lack of this perspective.

The five remaining sections are organized as follows: The first section describes a water supply system as a complex organizational system highlighting the attributes that demonstrate complexity, such as the concept of variety. The next section focuses on an appreciation of water supply from an open systems perspective, highlighting the interconnectivity of this infrastructure to other infrastructures. Next, the paper explores the characteristics of the system in focus and argues that the question of *exactly what this system is* should precede the classic risk assessment questions. The following section briefly reviews the literature on vulnerability assessment and concludes that the approaches are *ad hoc* checklists and lack systems-based perspective. This section also provides a working definition of vulnerability. Next, the paper addresses the Infrastructure Risk Analysis Model (IRAM) in its current form and suggests how the methodology could be improved by using systems principles and by addressing shortcomings in the vulnerability assessment step in the methodology. The final section summarizes the limitations of the framework and discusses potential uses of the methodology.

Water Supply as a Complex Organizational System

Complex organizational systems are defined as systems with tremendous variety that require large amounts of information to describe them. Variety (V_s) is measured by the number of possible different states for a system. Variety is a function of the number of system elements (n_e) and the number of possible different states (z_s) for each system element.

$$V_s = z_s^{n_e} \tag{1}$$

Table 1. Water system elements and function.

Element	Number of Elements	Function
Pumping Station	2	Transmit
Treatment Plant	2	
Reservoir	2	Store
Water Tank	4	Store
SCADA	1	Control and Monitor
Employees	18	Control, Manage, Maintain
Customers	8,000	Use

For example, in Table 1 the water system for a small community of 8,000 is described at the top level with only two states per element. There are

$2^3 * 4 * 1 * 18 * 8000 = 4,608$ entity relationships between the elements of the system. The variety then as a measure of complexity is $2^{4,608,000}$. If the customers are removed as elements in the system, the number of entity relationships in the system is 576. The variety, 2^{576}, remains a fantastically large number. (Inclusion or exclusion of customers as elements is dealt with later in the paper.) The actual number is not the dominant issue. The important point is the implication that the system's potential number of states is a surrogate measure for organizational complexity. The fact that people are included in the system suggests a complex system [Keating 2001].

Table 2 provides a matrix to assist in identifying a system as simple or complex[2]. Analysis indicates that the preponderance of descriptive evidence indicates a complex organizational system.

Table 2. Simple vs. complex organizational systems.

Characteristic	**Simple**	**Complex**	**Water Supply System**
Number of Elements	Small	Large	Large
Interactions between Elements	Few	Many	Many
Predetermined Attributes	Yes	No	No
Interaction Organization	Highly Organized	Loosely Organized	Loosely Organized
Laws Governing Behavior	Well-defined	Probabilistic	Probabilistic
System Evolution over Time	Not Evolved	Evolves	Evolves
Subsystems Pursue Own Goals	No	Yes (Purposeful)	Yes (Purposeful)
System Affected by Behavioral Influences	No	Yes	Yes
Predominantly Closed or Open to the Environment	Largely Closed	Largely Opened	Largely Opened

---

[2] Adapted from Jackson, 1991. *Systems Methodology for the Management Sciences.* In ENMA 815, Fall 2001.

Open Systems Analysis of Water Supply Systems

Open systems analysis can occur as hard (quantitative) or soft (qualitative) system studies. For example, sociotechnical systems (STS) analysis has more in common with soft systems study than with a hard systems approach. STS thinking has two fundamental main ideas. The first is that an organization must be regarded as an open system. The second idea is that people, the environment, and technology are interdependent. No component (social, technical, etc.) of the system is primary; rather, they are all equally important. STS recognizes that people form organizations and that an organization exists to carry out a primary task. A water utility's primary task is to provide safe drinking water to customers. In other words, the organization exists to serve a useful purpose beyond itself. It is clear that water supply systems exist in an open environment and largely are not closed. Von Bertalanffy [1969] defined an open system as "a system in exchange of matter, presenting import and export, build-up and breaking down of its material components." A tenet of open systems is that they are not in equilibrium but are in steady state. An open system such as a water supply system is homeostatic, meaning that it has the ability to maintain itself by transforming the environment. Other properties of open systems include the ability to self-regulate, evolve, reorganize, and achieve a desired state among different paths; this is referred to in the literature as the principle of equifinality. Last, an open system can use inputs from the environment to reduce entropy (the natural tendency of closed systems to gravitate towards chaos or destruction). Likewise, systems can be viewed as open cybernetic systems, as viable systems, socio-technical, and others. Critical infrastructures such as water supply systems may also be viewed as large-scale and interconnected to other critical infrastructures, touching many aspects of human life. Therefore, it seems reasonable to consider the utility of systems theory to help represent critical infrastructures as systems and to underpin vulnerability assessment.

The System in Focus

Characteristics commonly associated with systems perspectives are:
- ⇒ subsystems and hierarchies,
- ⇒ holism,
- ⇒ open systems,
- ⇒ boundaries,
- ⇒ negative entropy,
- ⇒ dynamic equilibrium,
- ⇒ in–trans–out,
- ⇒ feedback,
- ⇒ internal elaboration,
- ⇒ equifinality,
- ⇒ context,
- ⇒ emergent behavior,
- ⇒ interdependencies, and others.

Conceivably, there are more, but this list captures most of the important properties of the systems perspective. And each of these properties can be applied to water supply. A holistic view allows one to appreciate the scope and breadth of the system. An open system view acknowledges that water supply has many throughput boundaries, and the fact that it exists in an environment means that it requires energy to prevent it from becoming chaotic. There are inputs and outputs. For example, untreated water is transformed into clean water. Depending on the system boundaries, water is treated again as sewage and output water is returned to the environment. Feedback occurs in a cybernetic sense from the sensors at pumping stations, water tanks, and treatment plants. Water systems are inherently dependent upon electric power. This is just a short list of principles that show that water supply systems should properly be understood from a systems perspective before undergoing additional analysis. How can system vulnerability be conducted without a common understanding of the components, elements, functions, boundaries, and states of the system (among other factors)?

Defining Vulnerability and Assessment Approaches

Risk assessment methodologies are often employed to help understand what can go wrong, estimate the likelihood and the consequences [Kaplan 1997], and develop risk-mitigation strategies to counter risk and answer the traditional risk management questions [Haimes 1998]: What can be done? What are the tradeoffs? and What is the impact of current decisions on future options? The literature review indicates divergent views regarding vulnerability. Second, there is confusion among terms such as vulnerability, risk, hazard, assessment, analysis, etc. Buckle and Marsh [2000] contend that work must be done to clear up the definition of vulnerability with respect to risk. For example, Emergency Management Australia [EMA 1998] defines vulnerability as the degree of susceptibility and resilience to hazards of the community and environment. Likewise, the Emergency Management Australia [EMA 1998] glossary of terms confuses vulnerability analysis with hazard analysis or vulnerability assessment. The National Rural Water Association defines a vulnerability assessment as the identification of weaknesses in security, focusing on defined threats that could compromise its ability to provide a service [NRWA 2002]. Blaike et al. [1994] define vulnerability as "the characteristics of a person or group in terms of their capacity to anticipate, cope with, resist, and recover from the impact of a natural hazard." The National Oceanic and Atmospheric Administration [NOAA 2002] views vulnerability as the "susceptibility of resources to negative impacts from hazard events." Nilsson et al. [2001] contend that vulnerability is the collective result of risks and the ability of a society, local municipal authority, company, or organization to deal with and survive external and internal emergency situations. Gheorghe and Vamanu [2001] define vulnerability as the susceptibility and resilience/survivability of the community/system and its environment to hazards. In other words, vulnerability is a function of susceptibility, resilience, and the environment. The International Strategy for Disaster Reduction [ISDR 2001] defines vulnerability to disasters is "a status resulting from human action. It describes the degree to which a society is either threatened by or protected from the impact of natural hazards."

The literature review discloses that vulnerability assessments in use by risk practitioners, engineering managers, government, and private industry do not quantify system vulnerability. In addition, assessments appear to be checklists that lack methodological frameworks. See, for example NOAA [2002] or NRWA [2002]. One critical component of risk assessment methodology is determining the vulnerability of a system [Ezell et al. 2000a], but a theory of vulnerability seems to be missing from the literature. Research indicates a lack of theory and techniques to quantify system vulnerability.

For the purpose of this research, *critical infrastructure* is defined as "those physical and cyber-based systems essential to the minimum operations of the economy and government. They include, but are not limited to, telecommunications, energy, banking and finance, transportation, water systems, and emergency services, both governmental and private" [PDD63 1998].

The National Security Telecommunications Advisory Committee [NSTAC 1997] is credited with claiming that vulnerability is really a function of access and exposure, whereas Dictionary.com [2000] views vulnerability as "susceptibility to attack." If systems are exposed and accessible, then systems are susceptible; i.e., vulnerable to natural hazards as well as to willful intrusion, tampering, or terrorism. Vulnerability and risk appear to be coupled. Lowrance [1976] defines risk as a measure of the probability and severity of adverse effects. Vulnerability suggests susceptibility to risk [NOAA 2002].

Hierarchical Holographic Modeling (HHM) [Haimes 1981] has been used to identify sources of risk and indirectly imply systems vulnerabilities [Ezell et al. 2001]. Building upon NSTAC's [1997] ideas, the Infrastructure Risk Analysis Model (IRAM) [Ezell et al. 2000b] mathematically modeled vulnerability as simply a function of access and exposure. In water systems, exposure is equivalent to visibility. Water towers, treatment plants, and pump stations are examples of highly visible components. Through systems decomposition, vulnerability may be subjectively rank-ordered, based on exposure and access control. For example, a water system may be described in terms of components, elements, modes, or human interactions that satisfy an array of functions such as gathering, transmitting, and delivering. Within the current context of IRAM, a system is decomposed into components, and subjective *ad hoc* decisions are made on which source vulnerabilities to model. IRAM addressed vulnerability mathematically as a function of access α_i and exposure γ_i, where vulnerability of a component or subsystem is defined as $v_i = \alpha_i \gamma_i$, where α_i and γ_i are subjectively scaled $0 < \alpha_i < 1$ and $0 < \gamma_i < 1$. A low vulnerability score for a component is an advantage. The total vulnerability of the system is a simple summation of all systems vulnerability scores.

$$V = \sum_{i=1}^{n} v_i \qquad (2)$$

Figure 1 graphically depicts the steps in IRAM.

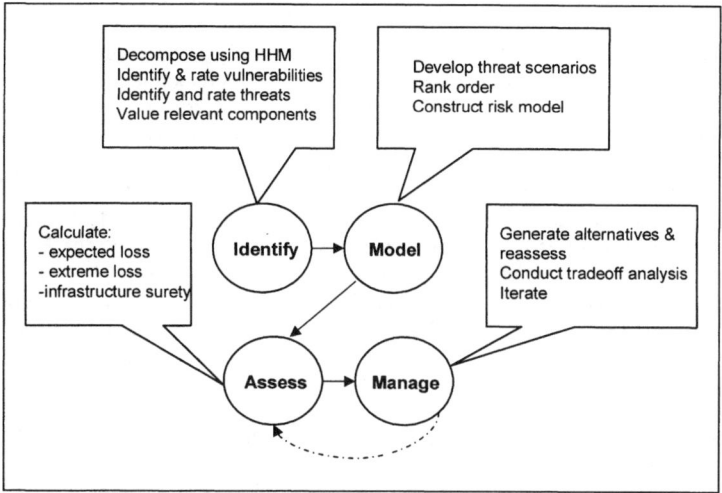

Figure 1. IRAM.

The vulnerability expression in Equation 2 above fails to account for the various differences in system size, complexity, number of components, etc. Also, it does not include many factors such as people, context, and others. The system boundaries for this model omit many items that touch the system. In addition, IRAM does not address the relative differences between systems. For instance, any large system's vulnerability score will always be larger than a smaller system's due to the number of components. There exist issues with the boundaries of critical infrastructures if such boundaries are simply around components from the perspectives of access and exposure. Systems theory may prove useful to bring agreement to definitions and systems views.

There have been a few attempts to address vulnerability either by defining or developing checklists within the context of its use. The following paragraphs summarize these:

The National Oceanic and Atmospheric Administration [NOAA 2002] views vulnerability as the "susceptibility of resources to negative impacts from hazard events." Hightower [2001] of Sandia National Laboratories presented a high-level discussion on water system vulnerability for local, state, and federal risk practitioners. However, Sandia has not shared its vulnerability assessment methodology to date. Ezell et al. [2000a and b] introduced the Infrastructure Risk Analysis Model and applied it to a small community's water supply system. It is the first documented attempt in the literature to quantify vulnerability. The National Security

Telecommunications Advisory Committee (NSTAC) [1997] presents a view stating that vulnerability may be viewed as access and exposure. Nilsson, Magnusson, Hallin, and Lenntorp [2001] investigated and presented methods suitable for analyzing and auditing municipal vulnerability. They proposed models which may make it possible to distribute financial support to municipalities for their work on reducing vulnerability in the most cost-effective way. Vulnerability exists as a result of a collection of risks and the ability of a society, local municipal authority, company, or organization to deal with and survive external and internal emergency situations.

Buckle [2000] says that vulnerability is a broad measure of the susceptibility to suffer loss or damage. The higher the resilience, the less likely damage may be, and recovery is likely to be faster and more effective. Conversely, the higher the vulnerability, the more exposure there is to loss and damage. Wenger et al. [2002] developed a handbook compiling risk analysis strategies used by eight countries for critical infrastructure protection. Organizations such as the Association of State Drinking Water Administrators, National Rural Water Association [NRWA 2002] have developed self-assessment vulnerability checklists, as has the American Water Works Association [AWWA 2002].

For the purposes of this paper:
⇒ exposed and accessible subsystems, elements, and components that are important (critical) to a system's purpose are vulnerable;
⇒ system vulnerability is characterized by the levels of criticality, exposure, accessibility, and resilience in the face of natural hazards, willful intrusion, tampering, or terrorism; and
⇒ the level of vulnerability inherent in a system is an indictment of the system's "health" in the face of adversity.

Methodology Development for Vulnerability Assessment

The philosophy of IRAM is based on the concept that one really cannot assess the system unless it is modeled and understood; hence, the emphasis on risk modeling, then assessment and management. IRAM is changed by addressing the preceding questions of assessment by identifying the system in focus. Next, the vulnerability index score derived from the subjective scales for access and exposure is replaced with a new philosophy of vulnerability borrowed from the quantitative definition of risk as a triplet of scenario likelihoods and consequences. As we know, risk is identified by the complete set of triplet questions which includes all the scenarios; that is, all the things that can go wrong. In the same way, we can define vulnerability as the set of triplet questions emerging from a given class of initiating events. Under this definition, vulnerability[3] is the risk resulting from a given set of initiating events.

---

[3]The ideas on vulnerability were developed in a personal conversation with Stan Kaplan at the Engineering Foundation Conference on Water Resources, Nov 9-11, 2002 in Santa Barbara, CA.

This implies that vulnerability by itself does not mean anything—there must be a vulnerability to something. Therefore, that "something" is a class of initiating events. Thus, the form of the assessment is a family of risk triplets (Figure 2).

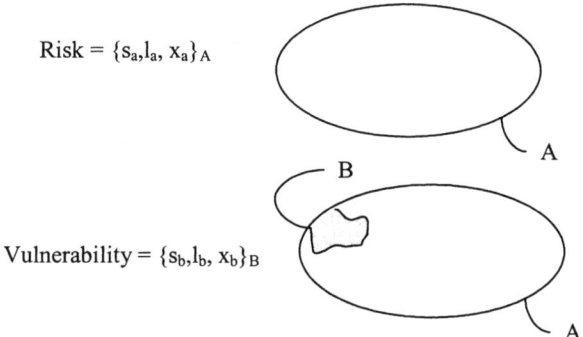

Risk = $\{s_a, l_a, x_a\}_A$

Vulnerability = $\{s_b, l_b, x_b\}_B$

Figure 2. Vulnerability as a family of risk triplets.

Building upon the quantitative definition of risk analysis [Kaplan 1997], consider the notion that the universal set of all risks to a system can be mathematically defined as a set of n triplets: R = f{scenario (s), likelihood (l), consequence (c)}. If there exists a list of n risks to a system, then a family or cluster of risk triplets can be viewed as a mathematical description of vulnerability to that system. For the purposes of this paper, *vulnerability is a measure of susceptibility to a family of risk triplets.* This definition captures the notion of "Susceptibility to what?" and "What are the consequences?" via the clustered triplet set as demonstrated in Figure 3. This figure shows the link from the universal set of all risk triplets, and the family of risk triplets that are true system vulnerabilities[4].

As an assessment methodology, IRAM is modified in Step 1: *Identify.* Vulnerability assessment, therefore, is now comprised of system identification, i.e., the system in focus and risk scenario development, with the family of risk triplets representing system vulnerability (Figure 4).

Assessment Limitations

The assessment framework provides a systematic approach that includes system representation and risk identification. The model is limited in that the techniques for developing scenarios are left to the practitioner. No attempt was made to address prescriptively how one ideates scenarios. There are several techniques well-documented in the literature (brainstorming, brain-writing, and dynamic

---

[4]This paper does not explore decision analysis techniques such as AHP or MAUT to derive the subset representing system vulnerability, as these techniques are well-documented in the literature.

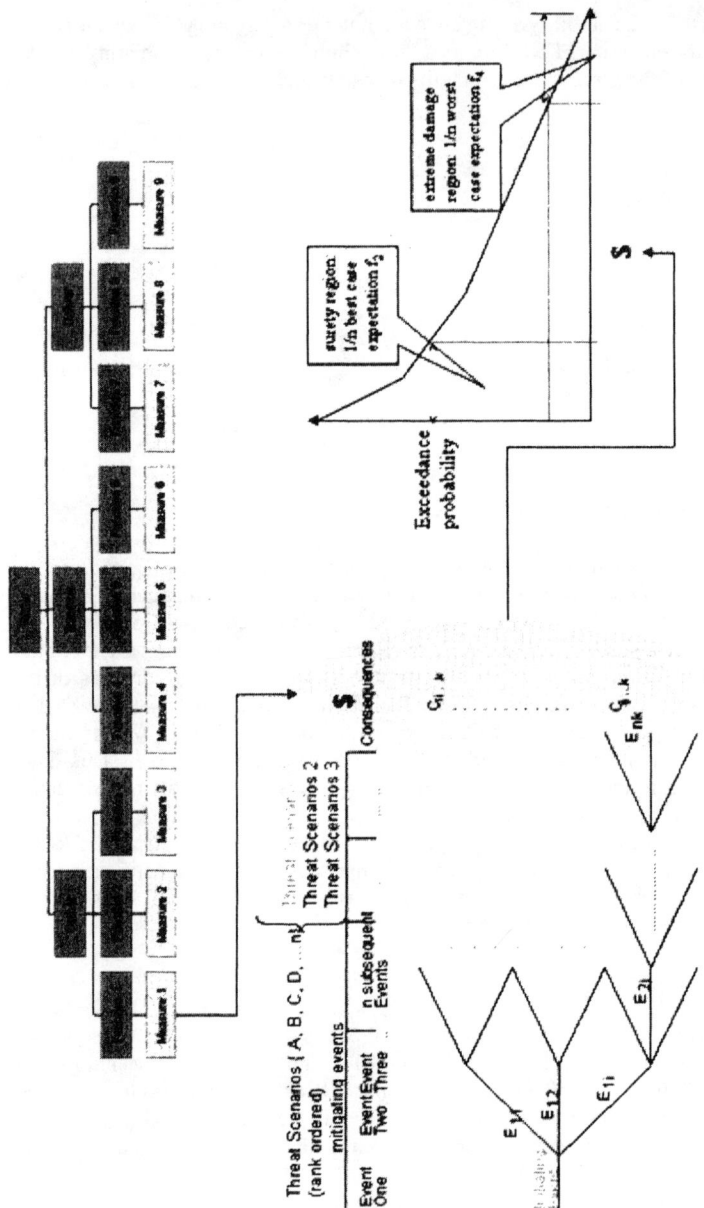

Figure 3. Linking the triplet from initiating event to evaluation measure.

confrontation, among others). Last, this version of the model provides no technique for deciding which risk triplet sets represent system vulnerability.

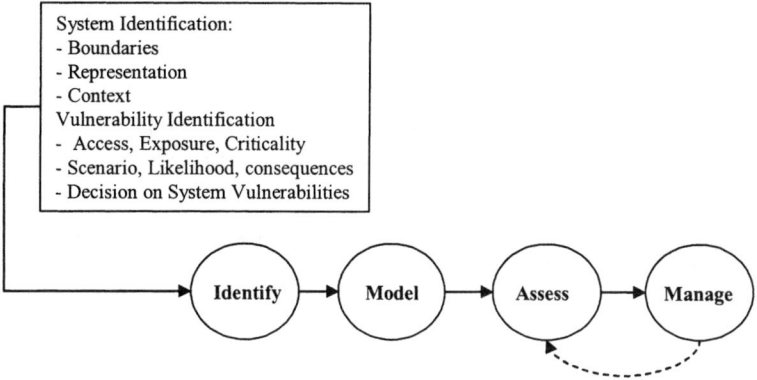

Figure 4. IRAM modified.

Conclusions and Future Work

This paper introduced a working definition of vulnerability and explained how the disciplines of quantitative risk analysis and systems theory might be used to define vulnerability. The methodology continues in development as a part of dissertation research. Future work includes:
⇒ the development of a theory of vulnerability where none exists;
⇒ the refinement of systems-based vulnerability assessment methodology; and
⇒ an improvement in how to quantify vulnerability for a critical infrastructure.

Practical implications of future work providing:
⇒ decisionmakers with a method to help them understand system vulnerability so that resources can be allocated in a meaningful way;
⇒ practitioners with the tools and references that will allow them to conduct their own analyses; and
⇒ military leaders with the ability to determine the vulnerability of systems that support operations.

References

AWWA (2002). *Emergency Planning for Water Utilities*. American Water Works Association, online at www.awwa.org/communications/offer/september.cfm.

Blaike, P., T. Cannon, I. Davis, and B. Wisner (1994). *At Risk: Natural Hazards, People's Vulnerability, and Disasters*. London, UK: Routledge.

Buckle, P. (2000). *Assessing Resilience and Vulnerability in the Context of Emergencies: Guidelines*. Victorian Government Publishing Service, online at www.anglia.ac.uk/geography/radix/ resources/buckle-guidelines.pdf.

Buckle, P. and G. Marsh (2000). "New approaches to assessing vulnerability and resilience." *Australian Journal of Emergency Management*, Winter, online at www.anglia.ac.uk/geography/radix/resources/buckle-marsh.pdf.

Dictionary.com (2000). http://www.dictionary.com.

EMA (2000). Emergency Management Australia, online at www.ema.gov.au.

Ezell, B.C., J.V. Farr, and I. Wiese (2000a). "Infrastructure risk analysis model." *Journal of Infrastructure Systems*, 6(3), 114-117.

Ezell, B.C., J.V. Farr, and I. Wiese (2000b). "An infrastructure risk analysis of a municipal water distribution system." *Journal of Infrastructure Systems*, 6(3), 118-122.

Ezell, B., Y.Y. Haimes, and J. Lambert (2001). "Risks of cyber attack to water utility supervisory control and data acquisition systems." *Military Operations Research*, 6(2), 23.

Gheorghe, A.V. and D.V. Vamanu (2001). *Fundamentals of Risk and Vulnerability Management QVA—Quantitative Vulnerability Assessment*. November 9, Zürich, online at www.isn.ethz.ch/crn/extended/workshop_zh /ppt/Gheorghe/tsld001.htm.

Haimes, Y.Y. (1981). "Hierarchical holographic modeling." *IEEE Transactions on Systems, Man, and Cybernetics*, 11(9), 606-617.

Haimes, Y.Y. (1998). *Risk Modeling, Assessment, and Management*. New York, NY: John Wiley and Sons.

Hightower, M. (2001). *Water Infrastructure Protection and Security—Emerging National Issues*. Distinguished member of the technical staff, energy, and critical infrastructure program, Sandia National Laboratories, Albuquerque, NM.

ISDR (2001). International Strategy for Disaster Reduction. Online at www.unisdr.org/unisdr/camp2001guide.htm.

Kaplan, S. (1997). "The words of risk analysis." *Risk Analysis,* 17(4), 407-417.

Keating, C. (2001). Class notes. Department of Engineering Management, ENMA 7/815, Class 4, Old Dominion University, Fall.

Lowrance, W.W. (1976). *Of Acceptable Risk.* Los Altos, CA: William Kaufmann.

Nilsson, J., S.E. Magnusson, P.O. Hallin, B. Lenntorp (2001). *Vulnerability Analysis and Auditing of Municipalities.* Lund University, Sweden, available on line at www.isn.ethz.ch/crn/basics/process/documents/vulnerability.pdf.

NOAA (2002). *Vulnerability Assessment.* National Oceanic and Atmospheric Administration, http://www.csc.noaa.gov/products/nchaz/htm/tut.htm.

NRWA (2002). *Security Vulnerability Self-Assessment Guide for Small Drinking Water Systems.* Association of State Drinking Water Administrators, National Rural Water Association, www.nrwa.com/downloads/SecurityAssessment.

NSTAC (1997). *Information Assurance Task Force Risk Assessment.* National Security Telecommunications Advisory Committee, available online at http://www.ncs.gov/n5_hp/reports/EPRA.html.

PDD63 (1998). *The Clinton Administration's Policy on Critical Infrastructure Protection: Presidential Decision Directive 63.* Available online at www.info-sec.com/ciao/paper598.pdf.

Von Bertalanffy, L. (1969). *General Systems Theory: Foundations, Development, Applications.* New York, NY: Braziler Publishing.

Wenger, A., J. Metzger, M. Dunn (2002). *International CIP Handbook: An Inventory of Protection Policies in Eight Countries.* Center for Security Studies and Conflict Research, Swiss Federal Institute of Technology, Zurich, Switzerland, September.

Demand-Reduction Input-Output (I-O) Analysis for Modeling Interconnectedness

Joost R. Santos[1] and Yacov Y. Haimes[2]

Abstract

The paper discusses the *demand-reduction I-O inoperability model* to analyze the economic impact of demand reductions on a system of interconnected infrastructures. The propagation of inoperability depends upon the degree of interdependency of one infrastructure on another. A case study comprising twelve critical interconnected infrastructure sectors is presented. The interdependency matrix for this case study is derived via a transformation of the national industry-by-industry transactions data as published by the Bureau of Economic Analysis. The paper highlights the use of *geographical* and *functional* decompositions to tailor the demand-reduction analysis to specific US regions, or to components of a large-scale infrastructure, respectively.

Introduction

Inoperability analysis in this paper connotes a process of studying how risks can propagate and proliferate through a system of interconnected infrastructures. *Perturbations* in the form of natural disasters, accidents, or willful attacks can set off a chain of cascading impacts, and thus risks, on interconnected infrastructures. A technical paper by the University of Virginia's Center for Risk Management of Engineering Systems (UVA-CRMES) asserts that the higher degrees of interdependencies exhibited by our critical infrastructures to date—due in part to their increasing reliance on modern technology—make them more vulnerable to willful attacks [UVA-CRMES 2002]. The September 11, 2001 attacks, for example, have

---

[1]Graduate student, Department of Systems and Information Engineering, and Center for Risk Management of Engineering Systems, University of Virginia; 434-924-3803; jrs8e@virginia.edu.
[2]Lawrence R. Quarles Professor of Systems and Information Engineering, and Director, Center for Risk Management of Engineering Systems, University of Virginia; 434-924-3803; haimes@.virginia.edu.

demonstrated the strong interdependence and interconnectedness in various infrastructures and sectors of the United States.

Input-output (I-O) analysis was developed by Leontief in the 1930s to present a framework for addressing the interconnectedness among various sectors of the economy [Leontief 1951a and b, 1966]. Utilizing this framework, Haimes and Jiang [2001] developed the Leontief-based I-O inoperability model to describe how the impact of perturbations can cascade through a system of interconnected infrastructures. In this model, the term *inoperability* connotes the level of the system's dysfunction, expressed as a percentage of the system's nominal level of operation. Furthermore, inoperability is interpreted as the degradation of a system's capacity (or supply) to deliver its intended output due to the physical impact of such perturbations. Thus, we refer to the Haimes and Jiang I-O inoperability model as either *physical-based* or *supply-based*.

Through the *demand-reduction inoperability I-O model* (or *demand-based model*, for brevity), the paper aims to complement and supplement the already-developed physical-based model. While the physical-based model quantifies inoperability in terms of degraded capacity to deliver the intended outputs, the demand-based model addresses the demand reductions that can potentially stem from perturbations. Logically, the demand reduction of a perturbed industry further renders adverse impacts on the operation of other dependent industries.

The remainder of the paper is organized as follows: Section 2 highlights the literature sources relevant to the usage of I-O analysis. Section 3 discusses the demand-based model and demonstrates it via a case study comprising 12 national critical interconnected infrastructures. Sections 4 and 5 provide an array of methods to further enhance the demand-based model. Specifically, these methods include geographical decomposition (Section 4), and functional decomposition (Section 5). Finally, the Section 6 epilogue highlights how the demand-based model is integrated into a higher-level I-O framework via the temporal modeling of inoperability.

Input-Output Models

I-O analysis is concerned with modeling interdependencies among a system of interconnected entities. These entities can be characterized in terms of industries—as described by the Leontief economic I-O model—or of critical infrastructures—as described by the physical-based inoperability I-O model.

Leontief economic I-O model. The I-O analysis was formally introduced by Wassily Leontief in the 1930s to present a framework for the study of economic equilibrium. Through the use of I-O tables, the model is capable of addressing the interconnectedness among various sectors of the economy [Horton 1995]. Leontief was awarded the 1973 Nobel Laureate in Economics for his seminal work on the development of the I-O method and for its application to important economic problems. Miller and Blair [1985] provide a comprehensive introduction to the model and its applications. Leontief's I-O model describes the equilibrium and dynamic behavior of both regional and national economies. Thus, it is a useful tool in economic decisionmaking processes used in many countries [Peterson 1991]. Recent frontiers in I-O analysis were compiled by Lahr and Dietzenbacher [2001].

The US Department of Commerce [1998] maintains various types of economic tables (or matrices) through its Bureau of Economic Analysis (BEA) division. The BEA is responsible for documenting the transactions among various industries in the US economy. The detailed national I-O accounts are composed of nearly 500 industries, organized according to the Standard Industry Classification (SIC) codes. For brevity, we focus only on describing the make (**V**) and use (**U**) matrices. The *make* matrix would show the dollar values of the different column commodities *produced* by the different row industries. The *use* matrix, on the other hand, would show the dollar values of the different row commodities *consumed* by the different column industries. The BEA data do not directly specify the I-O matrix representing the industry-by-industry transactions. This matrix, which is called the *industry-by-industry technical coefficient matrix* in Leontief parlance, shows the proportion of the ith industry's total output which serves as the input to the jth industry's production. To derive the industry-by-industry technical coefficient matrix, we first need to normalize the use and make matrices with respect to the total commodity output (**y**) and total industry output (**x**) vectors, respectively. The described operations will yield the normalized make (**D**) and the normalized use (**B**) matrices.

$$d_{ij} = \frac{v_{ij}}{y_j} \Leftrightarrow \mathbf{D} = \mathbf{V}[diag(\mathbf{y})]^{-1} \tag{1}$$

$$b_{ij} = \frac{u_{ij}}{x_j} \Leftrightarrow \mathbf{B} = \mathbf{U}[diag(\mathbf{x})]^{-1} \tag{2}$$

The operator $diag(\theta)$ in (1), (2), and later equations represents the resulting diagonal matrix constructed from a given vector θ, i.e.,

$$diag(\theta) = diag\begin{bmatrix} \theta_1 \\ \theta_2 \\ \vdots \\ \theta_n \end{bmatrix} = \begin{bmatrix} \theta_1 & 0 & \cdots & 0 \\ 0 & \theta_2 & \ddots & \vdots \\ \vdots & \ddots & \ddots & 0 \\ 0 & \cdots & 0 & \theta_n \end{bmatrix} \tag{3}$$

We use the notation **A** to refer to the industry-by-industry technical coefficient matrix. Miller and Blair [1985] provide the derivation for **A** to be the product of the normalized make and the normalized use matrices.

$$a_{ij} = \sum_k d_{ik} b_{kj} \Leftrightarrow \mathbf{A} = \mathbf{DB} \tag{4}$$

Defining **c** as the final demand vector for the industries, then the Leontief balance equation containing the industry-by-industry technical coefficient matrix (**A**) and the total industry output vector (**x**) is written as follows:

$$\mathbf{x} = \mathbf{Ax} + \mathbf{c} \quad (5)$$

Physical-based inoperability input-output model. Modern infrastructures to date generally manifest higher degrees of complexity and interconnectedness, due largely to their increasing reliance on information technology [Longstaff et al. 2000; Longstaff and Haimes 2002]. Thus, it is likely that the vulnerability of a given infrastructure could pose additional risks to other dependent infrastructures [Haimes 2002]. A first-generation physical-based inoperability I-O model (or physical-based model, for simplicity) was developed by Haimes and Jiang [2001] to describe how the impact of terrorist attacks can cascade through a system of interconnected infrastructures. Inoperability connotes degradation in the system's functionality (expressed as a percentage relative to the intended state of the system). The mathematical formulation of the physical-based model is as follows:

$$\mathbf{x} = \mathbf{Ax} + \mathbf{c} \quad (6)$$

However, note that the interpretation of the model parameters in (6) is fundamentally different from the Leontief model in (5). The "supply" and "demand" concepts in the Leontief economy model now assume different interpretations and have been inverted to some extent in the physical-based inoperability I-O model. Although the mathematical construct of the two models is similar, in Leontief's model, **x** and **c** represent commodities typically measured in dollar units. In the physical-based model, the vector **c** represents the *input* to the interconnected infrastructures—perturbations in the form of natural events, accidents, or willful attacks. The *output* is defined as the resulting vector **x** of inoperability of the different infrastructures, due to their connections to the perturbed infrastructure and to one another. The long-run inoperabilities of the interconnected infrastructures following an attack can be calculated using (6) provided that **A** is stable.

The inoperability vector (**x**) describes the degree of functionality of interconnected infrastructures. Thus, it takes on values between 0 and 1, where flawless operation corresponds to $\mathbf{x} = 0$ or $x_1 = x_2 = \cdots = x_n$. When this condition is in effect, the infrastructures are said to be at their *nominal* or *ground state*. A perturbation input **c** will cause a departure from the ground state. It can intuitively set off a chain of effects leading to higher-order inoperabilities—coined as *cascading effects* by Rinaldi [1997]. For example, a power infrastructure (the kth infrastructure) would initially lose 10% of its functionality due to an attack that delivers a perturbation (c_k) of 0.1. This defines the perturbation as the inoperability of the power infrastructure *right after* an attack. In addition, this inoperability propagated onto other power-dependent infrastructures will in turn cause other inoperabilities and ultimately perhaps additional inoperability in the power infrastructure itself. In general, we expect the long-run inoperability of an attacked infrastructure to increase from its post-attack value (i.e., the perturbation).

Demand-Reduction Inoperability Input-Output Model

Continued exploration of the prior work on I-O modeling (i.e., the physical-based model) indicates that no single model is capable of capturing the multiple visions, perspectives, or dimensions of a system. This philosophy is the basis for hierarchical holographic modeling (HHM)—a modeling schema for identifying and structuring multiple risk scenarios triggered by deviations from the system's "as planned" scenario [Haimes 1981, 1998; Kaplan et al. 2001]. Thus, while the already-developed physical-based inoperability I-O model analyzes the physical losses caused by natural and human-caused catastrophic events, it is necessary to consider other factors as well. Psychological factors, for one, have been shown to mirror the physical destruction delivered by such events. A comprehensive survey of the psychological effects of various types of disasters is documented by Norris et al. [2002]. Empirical studies such as those conducted by Susser et al. [2002] and Galea et al. [2002] specifically show the significance of the "fear factor" induced by the September 11, 2001 terrorist attacks. These papers suggest that fear can cause the public to reduce its demand for the goods/services produced by an attacked entity. Public apprehension of the safety of air transportation post 9/11, for example, caused a drastic reduction in the operations of airlines and of airline-dependent industries. Such retrenchments and changes in demand can have compelling economic repercussions (e.g., degraded production capacity) which add to the physical losses. The *demand-reduction inoperability I-O model* will be used to analyze the long-run adverse effects of demand degradation on the nominal operation levels of interdependent industries.

Model description. Central to demand-reduction inoperability I-O modeling is the analysis of how demand reduction can propagate from the directly attacked industry to others. As with the case of the physical-based model, the propagation of demand-based inoperability depends on an interdependency matrix—a matrix which describes the degree of coupling between infrastructures. By deriving the relationship between the demand-based I-O inoperability model and the original Leontief economic model, we establish a process of generating the interdependency matrix based on available economic data. The correspondence between these two models is given in UVA-CRMES [2002]:

$$q = A^* q + c^* \qquad (7)$$

where the variables are defined as follows:
- c^* is the vector of normalized degraded demand (i.e., nominal demand minus post-attack demand, divided by the nominal production);
- A^* is the interdependency matrix, whose elements are derived from the industry-by-industry technical coefficient matrix (A); and
- q is the vector of normalized production loss whose elements represent the ratio of unrealized production (i.e., nominal production minus post-attack production, divided by nominal production).

The normalized production loss (**q**) in the model is triggered by a terrorist-induced normalized degraded demand (**c***). We also refer to **q** as the *demand-based inoperability*. The interdependency matrix (**A***) in the demand-based I-O inoperability model can be generated based on a published industry-by-industry technical coefficient matrix. Thus, the demand-based I-O inoperability model can utilize the vast database available in the reports of the BEA. Through a transformation of the economic data collected and published by the agency, a sample demonstration of the demand-reduction inoperability I-O model is presented in the following section.

Sample implementation of the demand-reduction inoperability I-O model. The following example consists of the 12 representative industry sectors enumerated in Table 1. The data are obtained from US national I-O accounts released by the BEA [US Department of Commerce 1998]. These accounts contain the total industry outputs, denoted by the vector **x**$^\wedge$ in Table 2. Using BEA's *Make* and *Use* tables, the resulting industry-by-industry technical coefficient matrix (**A**) for the twelve industry sectors is presented in Table 3.

Table 1. Industry sectors selected for the model.

Index	SIC[†] Code	Description
1	7.0000	Coal
2	31.0101	Petroleum refining
3	65.0100	Railroads and related services
4	65.0301	Trucking and couriers
5	65.0400	Water transportation
6	65.0500	Air transportation
7	66.0100	Telephone and telegraph, communication services
8	68.0100	Electric services
9	68.0301	Water supply and sewerage systems
10	70.0100	Banking
11	72.0101	Hotels
12	74.0000	Eating and drinking places

[†]Standard Industry Classification

Table 2. Total industry outputs ($\mathbf{x}^\wedge$), in million $.

Industry	j=1	j=2	j=3	j=4	j=5	j=6	j=7	j=8	j=9	j=10	j=11	j=12
$x^\wedge_j$	26,917	132,281	35,588	157,105	32,440	94,141	180,317	170,896	3,715	268,591	52,407	280,708

Table 3. Industry-by-industry technical coefficient matrix (**A**).

Industry	j=1	j=2	j=3	j=4	j=5	j=6	j=7	j=8	j=9	j=10	j=11	j=12
i=1	0.1130	0.0000	0.0000	0.0000	0.0003	0.0000	0.0000	0.0917	0.0000	0.0000	0.0000	0.0000
i=2	0.0168	0.0618	0.0511	0.0458	0.0208	0.0930	0.0009	0.0120	0.0046	0.0009	0.0027	0.0023
i=3	0.0274	0.0016	0.0617	0.0018	0.0003	0.0006	0.0000	0.0263	0.0005	0.0001	0.0003	0.0011
i=4	0.0114	0.0042	0.0028	0.1569	0.0035	0.0028	0.0014	0.0038	0.0106	0.0110	0.0103	0.0096
i=5	0.0050	0.0042	0.0014	0.0012	0.1247	0.0016	0.0000	0.0041	0.0000	0.0001	0.0003	0.0003
i=6	0.0034	0.0004	0.0045	0.0036	0.0023	0.0614	0.0031	0.0022	0.0018	0.0036	0.0038	0.0030
i=7	0.0013	0.0011	0.0013	0.0129	0.0008	0.0128	0.1236	0.0017	0.0207	0.0092	0.0069	0.0035
i=8	0.0190	0.0093	0.0016	0.0046	0.0032	0.0027	0.0032	0.0001	0.0167	0.0052	0.0306	0.0195
i=9	0.0000	0.0001	0.0001	0.0000	0.0001	0.0001	0.0000	0.0000	0.0000	0.0000	0.0004	0.0001
i=10	0.0065	0.0066	0.0206	0.0066	0.0145	0.0065	0.0092	0.0124	0.0116	0.0474	0.0319	0.0077
i=11	0.0030	0.0007	0.0030	0.0038	0.0014	0.0032	0.0030	0.0024	0.0019	0.0031	0.0039	0.0029
i=12	0.0035	0.0017	0.0053	0.0045	0.0018	0.0194	0.0036	0.0028	0.0027	0.0036	0.0057	0.0150

Table 4. Interdependency matrix (A*).

Industry	j=1	j=2	j=3	j=4	j=5	j=6	j=7	j=8	j=9	j=10	j=11	j=12
i=1	0.1130	0.0002	0.0000	0.0000	0.0003	0.0000	0.0000	0.5822	0.0000	0.0001	0.0001	0.0002
i=2	0.0034	0.0618	0.0138	0.0544	0.0051	0.0662	0.0012	0.0154	0.0001	0.0019	0.0011	0.0050
i=3	0.0207	0.0059	0.0617	0.0080	0.0003	0.0015	0.0002	0.1264	0.0001	0.0008	0.0004	0.0084
i=4	0.0020	0.0035	0.0006	0.1569	0.0007	0.0017	0.0016	0.0041	0.0003	0.0187	0.0034	0.0172
i=5	0.0042	0.0170	0.0015	0.0059	0.1247	0.0048	0.0002	0.0215	0.0000	0.0009	0.0004	0.0022
i=6	0.0010	0.0005	0.0017	0.0060	0.0008	0.0614	0.0059	0.0039	0.0001	0.0104	0.0021	0.0091
i=7	0.0002	0.0008	0.0002	0.0113	0.0001	0.0067	0.1236	0.0016	0.0004	0.0137	0.0020	0.0055
i=8	0.0030	0.0072	0.0003	0.0042	0.0006	0.0015	0.0033	0.0001	0.0004	0.0081	0.0094	0.0320
i=9	0.0001	0.0021	0.0008	0.0016	0.0008	0.0017	0.0024	0.0011	0.0000	0.0033	0.0052	0.0104
i=10	0.0006	0.0033	0.0027	0.0038	0.0018	0.0023	0.0062	0.0079	0.0002	0.0474	0.0062	0.0081
i=11	0.0015	0.0019	0.0021	0.0114	0.0009	0.0057	0.0105	0.0077	0.0001	0.0158	0.0039	0.0156
i=12	0.0003	0.0008	0.0007	0.0025	0.0002	0.0065	0.0023	0.0017	0.0000	0.0034	0.0011	0.0150

Table 5. Demand-based inoperabilities (Row 2) and equivalent dollar losses (Row 3) resulting from a 20% degradation in airline demand.

Industry	j=1	j=2	j=3	j=4	j=5	j=6	j=7	j=8	j=9	j=10	j=11	j=12
q_j	0.0003	0.0151	0.0005	0.0006	0.0015	0.2131	0.0017	0.0005	0.0004	0.0006	0.0013	0.0014
Loss ($)	$9M	$2B	$19M	$86M	$48M	$20B	$301M	$86M	$2M	$162M	$69M	$401M

Equation (8) expresses the relationship between the interdependency matrix $\mathbf{A}^*$ and the Leontief industry-by-industry matrix $\mathbf{A}$. UVA-CRMES [2002] provides the derivation for this relationship. Thus, plugging-in the entries of Tables 2 and 3 to (8) will yield the interdependency matrix shown in Table 4.

$$\mathbf{A}^* = [diag(\hat{\mathbf{x}})]^{-1} \mathbf{A}[diag(\hat{\mathbf{x}})] \Leftrightarrow a_{ij}^* = a_{ij}\left(\frac{\hat{x}_j}{\hat{x}_i}\right) \tag{8}$$

Now, suppose the airline industry experiences a terrorist-induced normalized degraded demand of 20% (i.e., this corresponds to a $c_6^* = 0.2$ and $c_i^* = 0.2$ for all $i \neq 6$). The resulting demand-based inoperabilities (**q**) due to interdependencies can be calculated using $\mathbf{q} = \mathbf{A}^*\mathbf{q} + \mathbf{c}^*$. These inoperabilities along with their equivalent dollar values are summarized in Table 5. The dollar equivalents are calculated by multiplying the inoperability of an industry i (i.e., q_i in Table 5) by its respective total industry output (i.e., $\hat{x}_i$ in Table 2).

The following can be observed from Table 5:
- Due to interdependencies, a demand reduction of 0.2 for the airline industry causes other industries to be inoperable as well.
- In terms of inoperability values (row 2 of Table 5), the next five most-affected industries are petroleum refining ($j=2$), telephone and communication services ($j=7$), water transportation ($j=5$), eating and drinking places ($j=12$), and hotels ($j=11$).
- In terms of actual dollar losses (row 3 of Table 5), the next five most-affected industries are petroleum refining ($j=2$), eating and drinking places ($j=12$), telephone and communication services ($j=7$), banking ($j=10$), and trucking and couriers ($j=4$).
- Although the inoperability of water transportation ($j=5$) is greater than that of banking ($j=10$), (i.e., 0.011 vs. 0.002, respectively), the dollar losses for water transportation are lower than for banking ($48M vs. $162M). *This suggests that the calculated inoperabilities and dollar losses are both important metrics in evaluating the impact of a demand reduction.* This is evident from the significantly different production scales of the aforementioned industries;
- Due to the dependency of the airline industry within itself (i.e., intra-dependency) and with other industries (interdependency), a demand reduction of 0.2 can ultimately reduce its "operability" by 0.2131.

Geographical Decomposition

Geographical decomposition enables a more focused and thus accurate analysis of infrastructure interdependencies for desired regions in the United States. Miller et al. [1989] and Lahr et al. [2001] discuss the validity of "closing" the I-O analysis to a particular region (i.e., a single regional I-O framework as opposed to multiregional) since interregional feedbacks empirically are found to be "small." The Regional I-O Multiplier System (RIMS-II) division of the US Department of Commerce is the agency responsible for releasing regional multipliers for various regions in the United States. Empirical tests suggest that regional multipliers can be used as surrogates for time-consuming and expensive surveys without compromising accuracy [US Department of Commerce 1997]. Given the national I-O accounts along with the *location quotients* (l_i), analysts can convert and customize the national data to more appropriate regional data.

$$l_i = \frac{\hat{x}_i^R / \hat{x}^R}{\hat{x}_i / \hat{x}} \qquad (9)$$

where: $\hat{x}_i^R$ regional output for the ith industry
$\hat{x}^R$ total regional output for the regional-level industries
$\hat{x}_i$ national output for the ith industry
$\hat{x}$ total national output for the national-level industries

Thus, the regional industry-by-industry technical coefficient matrix ($\mathbf{A}^R = \{a_{ij}^R\}$) can be calculated given the national industry-by-industry technical coefficient matrix ($\mathbf{A} = \{a_{ij}\}$) and the location quotients for the region R of interest.

$$a_{ij}^R = \begin{cases} a_{ij}(l_i) & l_i < 1 \\ a_{ij} & l_i \geq 1 \end{cases} \qquad (10)$$

Alternatively, when Σ is used to denote a unity vector, (10) can be written as follows:

$$\mathbf{A}^R = diag[Min(\mathbf{1},\Sigma)]\mathbf{A} \Leftrightarrow a_{ij}^R = Min(l_i,1)a_{ij} \qquad (11)$$

Functional Decomposition

Functional decomposition is commonly used to represent a large-scale system broken down into its various sub-components. It can be extremely valuable because it adds detail to the analysis [Haimes 1977; Haimes et al. 1990]. Functional decomposition involves dissecting a system according to the functions it performs. It avoids over-aggregation and therefore yields more realistic and accurate results. Most large-scale systems are composed of families of sub-systems operating in a hierarchical manner. In studying a large-scale organization, we consider not only the relationships among sub-systems at a specific level, but also their interactions with different sub-systems at different levels in the hierarchy. The operation of lower-level sub-systems is directly influenced by higher level sub-systems. On the other hand, the fulfillment of the higher-level goals depends on the lower-level performance. Thus, functional decomposition within a coordinated hierarchical framework is important to ensure that decisions at different subsystem levels will not conflict, and that the overall goal of the organization is realized successfully. When considering the transportation system, for instance, one may naturally decompose its scope to highlight its sub-systems, such as rail, highway, air, and water transportation. Figure 1—a BEA industry block—shows a sample of how industries can be disaggregated into their corresponding 6-digit Standard Industry Classification (SIC) sub-industries.

Epilogue

This paper presents a demand-reduction I-O model to assess terrorist-caused demand-based inoperabilities and their impacts on interdependent industries. A national-level case study comprising 12 major industries demonstrates the workings of the model. Two types of decomposition methods—*geographical* and *functional*—enhance the level of detail and resolution in the demand-reduction I-O model.

An initiative to develop a holistic I-O inoperability framework is already in progress. With this holistic framework depicted in Figure 2, we aim to pinpoint the roles of risk assessment [Kaplan and Garrick 1981; Kaplan 1997] and risk management [Haimes 1991, 1998] into the integrated physical-based and demand-based models. To further establish the linkage between these models, we also continue to study the temporal evolution of inoperability. Several time-frames, or *regimes,* exhibit different features of interdependencies following an attack. After an attack, the nature and extent of sector interactions will vary from time-frame to time-frame [UVA-CRMES 2002]. Therefore, in the temporal evolution study, the metrics of outcomes will be allowed to vary in the different time-frames. Within each time-frame, the holistic inoperability I-O risk model can describe a conceptual situation of equilibrium.

Before equilibrium is reached, the system will have evolved into a distinct and new frame of interactions. Both physical and psychological considerations will be accounted for in analyzing the long-run adverse economic impacts on the nominal operation levels of the interconnected infrastructures. A sample of several time-frames that will be addressed by the holistic I-O inoperability framework is presented in Figure 3.

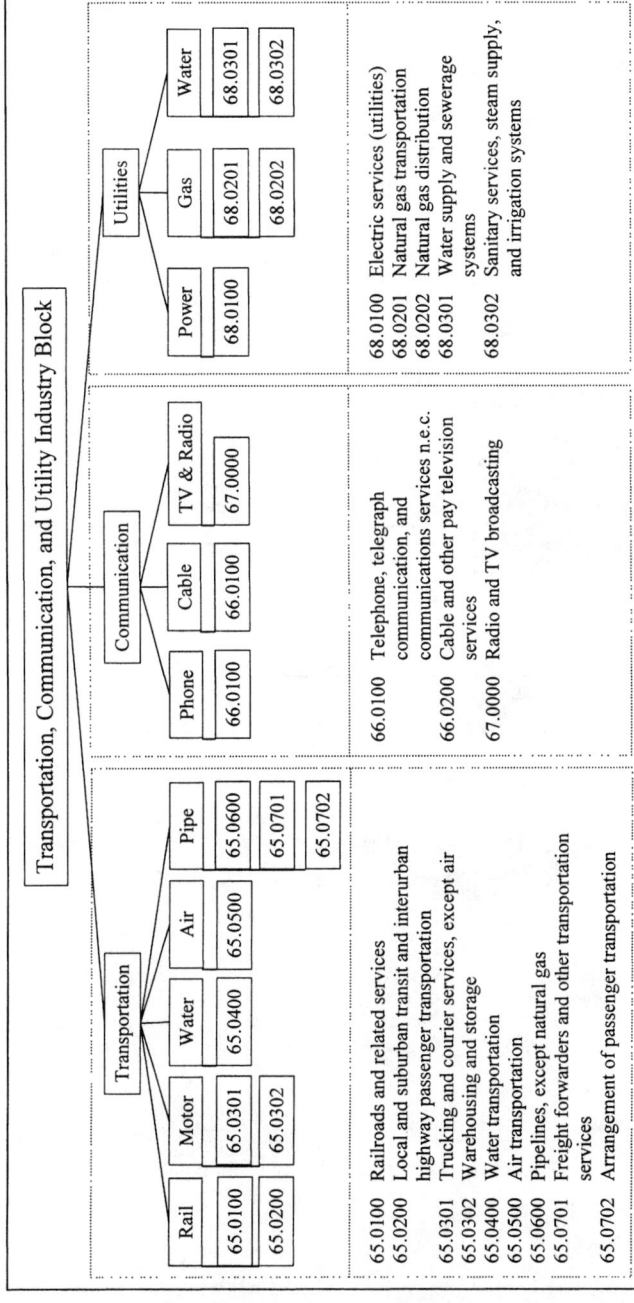

Figure 1. Sample of BEA industry decomposition.

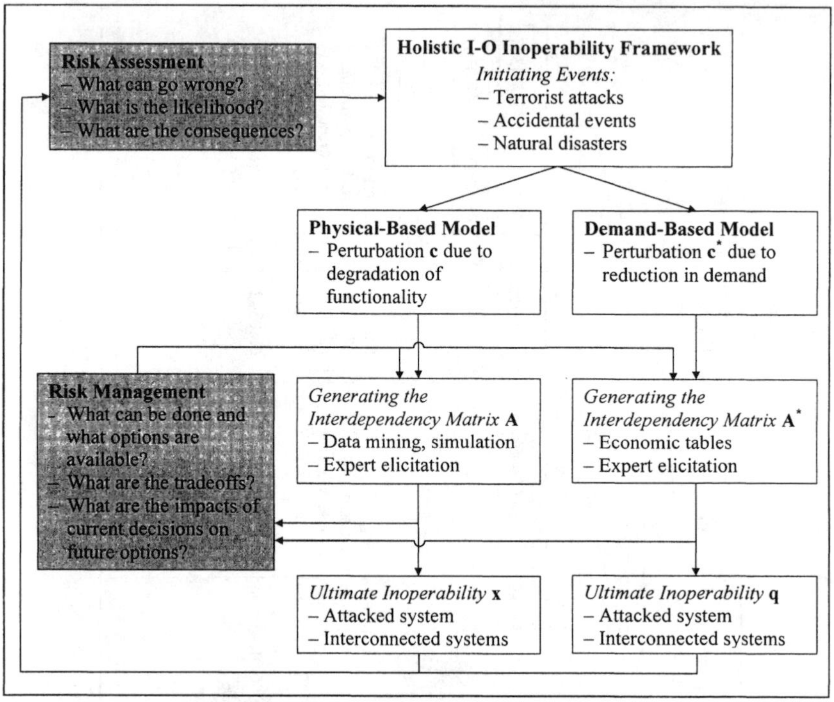

Figure 2. Holistic I-O inoperability framework.

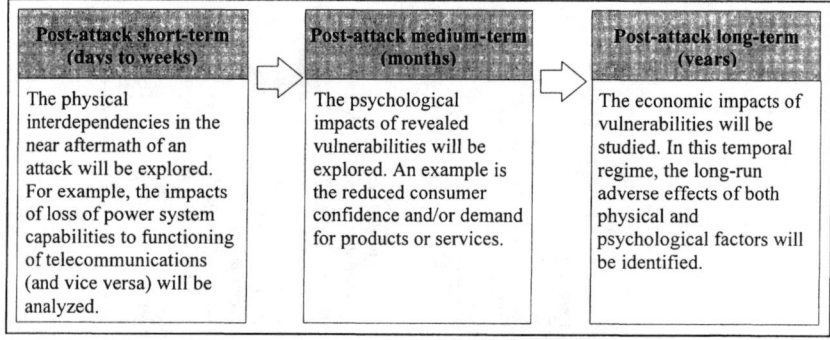

Figure 3. Temporal modeling of inoperability.

References

Galea, S., J. Ahern, H. Resnick, D. Kilpatrick, M. Bucuvalas, J. Gold, and D. Vlahov (2002). "Psychological sequelae of the September 11 terrorist attacks in New York City." *The New England Journal of Medicine* 346(13), 982-987.

Haimes, Y.Y. (1977). *Hierarchical Analyses of Water Resources Systems: Modeling and Optimization of Large-Scale Systems*. New York, NY: McGraw-Hill Book Company.

Haimes, Y.Y. (1981). "Hierarchical Holographic Modeling." *IEEE Transactions on Systems, Man, and Cybernetics* 11(9), 606-617.

Haimes, Y.Y. (1991). "Total risk management." *Risk Analysis* 11(2), 169-171.

Haimes, Y.Y. (1998). *Risk Modeling, Assessment, and Management*. New York, NY: John Wiley and Sons.

Haimes, Y.Y. (2002). "A roadmap for modeling the risks of terrorism to the homeland." *Journal of Infrastructure Systems* 8(2), 35-41.

Haimes, Y.Y. and P. Jiang (2001). "Leontief-based model of risk in complex interconnected infrastructures." *ASCE Journal of Infrastructure Systems* 7(1), 1-12.

Haimes, Y.Y., K. Tarvainen, T. Shima, and J. Thadathil (1990). *Hierarchical Multiobjective Analysis of Large Scale Systems*. New York, NY: Hemisphere Publishing.

Horton, G.A. (1995). *Input-output Models and Economic Impact Analysis: An Overview of Methodology, Economic Impact Multipliers, and Comparisons to Economic Forecast Models, with a Glossary of Terminology and Selected References*. Reno, NV: Business and Economic Research Associates.

Kaplan, S. (1997). "The words of risk analysis." *Risk Analysis* 17(4), 407-417.

Kaplan, S. and B.J. Garrick (1981). "On the quantitative definition of risk." *Risk Analysis* 1(1), 11-27.

Kaplan, S., Y.Y. Haimes, J. Garrick (2001). "Fitting hierarchical holographic modeling (HHM) into the theory of scenario structuring and a refinement to the quantitative definition of risk." *Risk Analysis* 21(5), 807-819.

Lahr, M.L. and E. Dietzenbacher (Eds.) (2001). *Input-Output Analysis: Frontiers and Extensions*. New York, NY: Palgrave.

Leontief, W.W. (1951a). "Input-output economics." *Scientific American* 185(4).

Leontief, W.W. (1951b). *The Structure of the American Economy, 1919-1939*, Second Edition. New York, NY: Oxford University Press.

Leontief, W.W. (1966). *Input-Output Economics*. New York, NY: Oxford University Press.

Longstaff, T.A. and Y.Y. Haimes (2002). "A Holistic Roadmap for Survivable Infrastructure Systems." *IEEE Transactions on Systems, Man, and Cybernetics* 32(2), 260-268.

Longstaff, T.A., C. Chittister, R. Pethia, and Y.Y. Haimes (2000). "Are we forgetting the risks of information technology?" *Computer* 33(33), 43-51.

Miller, R.E. and P.D. Blair (1985). *Input-Output Analysis: Foundations and Extensions.* Englewood Cliffs, NJ: Prentice-Hall, Inc.

Miller, R.E., K.R. Polenske, and A.Z. Rose (1989). *Frontiers of Input-Output Analysis*. New York, NY: Oxford University Press.

Norris, F.H., C.M. Byrne, E. Diaz, and K. Kaniasty (2002). "The range, magnitude, and duration of effects of natural and human-caused disasters: a review of empirical literature." *A National Center for Post-Traumatic Stress Disorder (PTSD) Fact Sheet*, http://www.ncptsd.org/facts/disasters/fs_range.html, May 31, 2002.

Peterson, W. (1991). *Advances in Input-Output Analysis: Technology, Planning, and Development*. New York, NY: Oxford University Press.

Rinaldi, S.M. (1997). "Complexity theory and airpower: a new paradigm for airpower in the 21st century." In *Complexity, Global Politics, and National Security*, David S. Alberts and Thomas J. Czerwinski (Eds.). Washington, DC: National Defense University Press.

Susser, E.S., D.B. Herman, and B. Aaron (2002). "Combating the terror of terrorism." *Scientific American* 287(2), 70-77.

US Department of Commerce, Bureau of Economic Analysis (1997). *Regional Multipliers: A User Handbook for the Regional Input-Output Modeling System (RIMS II)*. Washington, DC: US Government Printing Office.

US Department of Commerce, Bureau of Economic Analysis (1998). *Benchmark Input-Output Accounts of the United States, 1992*. Washington, DC: US Government Printing Office.

UVA-CRMES (2002). *Technical Paper: Extensions of the Leontief-Based Input-Output Inoperability Model.* University of Virginia Center for Risk Management of Engineering Systems, Charlottesville, VA.

Belief Systems and Reducing Risks from Terrorism

Robert E. O'Connor[1]

Abstract

Threats to water resources from terrorists are fundamentally different from threats from hurricanes and other natural phenomena because terrorists can choose their targets. As a result, effective approaches to disaster prevention should focus more on identifying and apprehending terrorists rather than reducing the vulnerability of water systems. Certain popular belief systems, however, are inconsistent with expanded government use of methods designed to spot terrorists, such as sensors, surveillance, and data mining. Progress in reducing risks requires wise organizational responses to these widespread concerns.

Introduction

Risk analysts use a variety of methods to assess risks of extreme events [Bier et al. 1999].[2] These methods include, for example, Bayesian techniques, maximum entropy distributions, extreme value theory, modeling, and decomposition. Both scholars and practitioners have used these as well as traditional statistical approaches to analyze extreme events. Terrorist attacks, however, are fundamentally different from natural disasters. The use of risk-analytic methods in the face of terrorism leads to a conclusion that the most effective approach to preventing such attacks is to identify and apprehend terrorists, not to focus on protecting targets.

This paper argues that policies designed to prevent water resource disasters by increasing surveillance, sensors, and data mining are likely to generate widespread public opposition. However, understanding the sources of public opposition in particular belief systems will help policymakers respond responsibly and effectively to public concerns.

---

[1] Director of the Decision, Risk, and Management Sciences Program at the National Science Foundation, 4201 Wilson Boulevard., Suite 995, Arlington, VA 22230; 703-292-7263; roconnor@nsf.gov.
[2] Extreme events are both infrequent and severe in their consequences. They include both natural events such as floods (admittedly, poorly designed/built environments may exacerbate losses) and deliberate human attacks.

The Policy Problem

Several presentations and numerous *ad hoc* discussions at the Engineering Foundation conference highlighted the vulnerability of water resources to terrorist attacks. There was less certainty, however, in identifying appropriate prevention measures. A National Academy Sciences panel [NRC 2002] concludes:

> Prevention measures may be sought at five points in the terrorist process: (a) long-term efforts to modify the demographic, economic, political, and cultural background of terrorism; (b) prevention at the source, that is, by seeking out and disrupting terrorist activities—in the present case, in staging areas in the several countries that knowingly harbor terrorists and to some extent in countries that do not wish them there; (c) prevention at the end of the line, by erecting defenses at the locus of known or conceivable targets, such as dams, public buildings, mass assemblages of people, and so on; (d) along the way between source and event, by controlling the movements of people and weapons at the national borders and other points of entry; and (e) after an attack, by having in place a response-and-recovery apparatus that will minimize its effects.

Each of the five prevention measures is laudable, yet leaves water resources vulnerable. Addressing the root causes of terrorism is important, but this approach does little to prevent attacks in the foreseeable future. Disrupting terrorist activities abroad makes sense, but is unlikely to eliminate most potential terrorists. Strengthening targets also makes sense and has an ancillary benefit of reducing damage from vandals, but society will never have the resources to protect every facility. Improving border security also seems sensible, but the task is technically difficult and expensive, and does not address domestic terrorists and foreign terrorists already in the country. Improving rescue and recovery planning would reduce losses, but is not preventative. What is missing from the list is the approach of identifying and apprehending terrorists in the country before they act.

Stan Kaplan [Kaplan 2003] uses game theory to show that terrorists have an enormously significant advantage in their ability to pick their targets. If one community water system makes great strides toward protecting itself from terrorist attack, one consequence is to increase the comparative appeal of other community water systems or other critical infrastructure facilities as targets. Kaplan does not argue that facilities should not strengthen their security, but shows that the only way to protect water systems from terrorists is to find the terrorists before they act.

Finding terrorists before they act means that government agencies will increase their use of surveillance activities and sensors. Finding terrorists also means advanced data mining of the flood of data Americans generate as we live our lives. The idea is to identify suspicious patterns from the data. The Defense Advance Research Projects Agency of the Department of Defense has begun such a program: *Total Information Awareness*. Surveillance, sensors, and data mining, intended to reduce risks from terrorism, nevertheless do not mesh easily with our liberal political heritage.

Three American Belief Systems

Louis Hartz in his classic, *The Liberal Tradition in America* [1955], declared that the key to understanding Americans is our origin as a nation in flight from the feudal and clerical oppressions of the Old World. He argued that, with only a few aberrations (e.g., parts of the ante-bellum South), the dominant culture has always embraced the essential liberal belief in individual freedom. Most Americans still instinctively recoil at any argument that suggests that someone (e.g., the government, a religious leader) has the right to tell us how we must live our lives. Instead, we still accept the liberal epistemology (whether we call ourselves conservatives or liberals) that each of us must be free to pursue whatever life goals we choose, so long as we do not interfere with others, because nobody can know the cosmic normative "Truth." The vast majority of us are liberals in the sense that we embrace this premise, yet three distinct belief systems have evolved within this liberalism.[3]

The three belief systems are described below as ideal types. In practice, individual Americans often hold views consistent with more than one belief system. For example, there is nothing inherently contradictory in holding a number of attitudes consistent with both moralistic conservatism and modern liberalism. Nevertheless, in order to consider the implications of common belief systems for the acceptance of specific approaches to reducing terrorism, it is useful to present different dimensions of classical liberalism as separate belief systems.

Libertarian conservatism. The threat to freedom in the minds of libertarian conservatives is government. Their intellectual forebears are the early philosophies of John Stuart Mill and Adam Smith. In *On Liberty* [1859], Mill wrote that the state has no right to suppress even the most heinous ideas. In *The Wealth of Nations* [1776], Smith argued that markets unrestrained by government intervention produce societal decisions that are responsive to public demands as if an "invisible hand" were manipulating prices and supplies. Freedom is freedom from government.

The proper role for government in the libertarian conservative heaven is a minimal one, essentially that of defending the nation's borders and performing domestic police powers. Libertarian conservatives posit that even the best-intentioned government interventions usually produce more harm than good, so governments should do as little as possible. At the extreme, libertarian conservatives view the state as a dangerous leviathan that seeks to increase its power at the expense of the freedom of individuals.

Moralistic conservatism. The threat to freedom in the minds of moralistic conservatives is anarchy and its companion, moral decay. Their intellectual forebear is Thomas Hobbes. In *Leviathan* [1651] Hobbes painted a bleak picture of the lives of people living without strong government. For Hobbes, life in such a state of nature for individuals would be "solitary, poor, nasty, brutish, and short" because people

---

[3] Although not all scholars agree that the American liberal belief system breaks neatly into three variants, this account is consistent with mainstream views. See, for example, McCloskey and Zaller [1984] and Gerzon [1996].

naturally compete. The solution for Hobbes was monarchy because monarchy solves the problem of succession without bloodshed and provides for authoritarian government to maintain peace and civility. Most modern moralistic conservatives differ from Hobbes in that they do not favor monarchy, but agree with Hobbes that strong government is needed to provide stability. Freedom is freedom from fear that things are falling apart.

The proper role of government in the moralistic conservative heaven is a stern and powerful paternalism that maintains security through strengthening community norms and promoting moral behavior. Moralistic conservatives are concerned that the glue that holds society together may be weakening, so the state should foster shared communal identities as well as moral behavior. At the extreme, some moralistic conservatives favor state support for majoritarian religious institutions as a means of promoting common values. For moralistic conservatives, the worst thing the government can do is to promote multiculturalism.

Some moralistic conservatives reject both traditional and modern liberalism in that they want the state to enforce lifestyle norms that would repress minority lifestyles and norms. However, most moralistic conservatives still fall under traditional notions of liberalism because they do not want the government telling people how they must live their lives. Nevertheless, moralistic conservatives are willing to give the government greater leeway in promoting common symbols than would either libertarian conservatives or modern liberals.

Modern liberalism. The threat to freedom in the minds of modern liberals is the lack of opportunity for individuals to make meaningful choices for their lives. Their intellectual forebear is Thomas Hill Green, a nineteenth-century British philosopher who coined the term "positive freedom." Green wrote [1880] that the earlier definition of liberalism by John Stuart Mill [1859] was essentially negative in that it is merely freedom from government coercion. Green agreed with Mill that the government should not censor writings, but found the Mill definition insufficient. Freedom for Green and modern liberals includes the notion of positive freedom, that in order to make the right to choose meaningful, society should ensure that individuals have the basic resources (e.g., food, shelter, education, medical care). Modern liberals have extended Green's thinking to include a healthy environment, broadly interpreted. For this group, freedom means that everyone will have the minimal conditions of human well-being, the community will provide a good place in which to live, and everyone will receive equal treatment under law.

The proper role of government in the modern liberal heaven is to ensure that everyone has the basic resources to make meaningful choices and that the common belief systems flourish. The government is neither the minimal presence of the libertarians nor the disciplinarian of the moralists. Modern liberals argue that a free society ensures that every child attends an adequate school, no family goes hungry because of poverty, and everyone has adequate medical care and shelter. A free society provides basic opportunities for all and equal treatment under the law. The government does not tell people how they should live their lives; rather, it acts as the provider of last resort to ensure meaningful choices, while promoting and preserving a healthy environment—for business, health, recreation—for all.

Belief Systems and Support for Hunting Terrorists

Libertarian conservatives are inherently wary of projects such as Total Information Awareness. They care deeply about privacy and do not trust the government not to behave badly with the information it collects and analyzes. Consistent with their conservative libertarian views, former Congressmen Dick Armey and Bob Barr have denounced the program, as have groups such as the American Conservative Union, Americans for Tax Reform, the Cato Institute, and the Free Congress Foundation.

In contrast, moralistic conservatives should support government efforts to find terrorists regardless of privacy issues. As expected, a web search fails to identify statements by moralistic conservatives critical of the Total Information Awareness program or similar government efforts to reduce risks from terrorism.

Modern liberals are skeptical toward the Total Information Awareness program and similar efforts because of their support for the equitable treatment of individuals and their concern that such programs might be used to thwart progressive policies.

The belief system of modern liberals includes the axiom that government should treat individuals the same regardless of their ethnicity, religious heritage, race, or sex. Government practices based on standard statistical analyses are consistent with this. Practices based on Bayesian methods, for example, may violate this axiom. For example, a program that singles out males of Middle Eastern ancestry for additional screening at airports violates it. Modern liberals assume that government programs of surveillance, sensors, and data mining will use factors such as ethnicity, religious heritage, race, and sex to narrow its searches. Moralistic conservatives see such narrowing as reasonable behavior designed to improve the quality of the search for terrorists. Modern liberals are not so sure that they can trust intelligence agencies, which leads to the second reason they tend to be skeptical. They perceive that government intelligence agencies have a long record of interfering with the lives of progressive leaders. J. Edgar Hoover is dead, but his perceived efforts to discredit Martin Luther King, Jr. remain a powerful memory. The appointment of John Poindexter of Iran-*contra* notoriety to head the Total Information Awareness program feeds concerns that the government will not be adequately sensitive to individual rights.

Conclusions and Discussion

Terrorism is a real threat, but Americans do not want to sacrifice their own freedom in an effort to reduce risks. The following six conclusions flow from the previous description of belief systems in the context of risks to water systems.

- Bayesian methods and data mining may well be wise and just at the aggregate level, yet seem unjust at the individual level. The American liberal political culture raises obstacles to the use of these methods because of our focus on individualism and individual rights. One indication of the difficulty with implementing these methods in a liberal society may be their absence from the list of potential preventive measures developed by a distinguished panel at the National Academy of Science.

- Technical arguments with game-theory assumptions, regardless of how sound the arguments may be, will not convince people who are skeptical of the value of increasing surveillance, sensors, and data mining. Opposition stems from tenets related to deeply-held belief systems, not from different benefit/cost calculations. The concern is not whether the proposed programs will work, but how they might work and be used.
- The dispute is not along a single dimension between political leftists and rightists, or twentieth-century conservatives and liberals. The sharpest disputes are probably between libertarian conservatives and moralistic conservatives. Therefore, efforts to link arguments to standard symbols of the left or the right are not likely to be effective.
- Risk communications that demonize proponents or opponents of programs such as Total Information Awareness are not likely to contribute to wise policy resolutions. Terrorist attacks are horrible and there is a strong logical argument that we would be wise to focus more energy on identifying and capturing terrorists in our country. Nevertheless, belittling the concerns of libertarian conservatives and modern liberals is not constructive. Simply asserting that they need to trust the intelligence community is insulting. A review of the history of the intelligence community suggests that critics have a reasonable basis for their concerns.
- Effective program oversight is the key to acceptance. Proponents of increased surveillance, sensors, and data mining assert that these activities can move forward without invading the privacy of anyone except terrorists. Most critics disagree. The implementation of strong oversight changes the discussion from general concerns associated with belief systems to narrower issues of oversight organization, planning, and sufficiency. People do not change their belief systems quickly, but they do change their opinions of specific programs as new information comes to light.

Capturing terrorists may be largely a technical problem, but the most effective methods will not go forward easily or quickly if they seem inconsistent with the belief systems of millions of Americans.

Acknowledgments

This paper emerged from discussions at the November conference. I particularly appreciate the leadership of Yacov Haimes in fostering the conference's tradition of open, stimulating exchanges. The views expressed in the paper are not necessarily those of the National Science Foundation or anyone else who knows me.

References

Bier, V., Y. Haimes, J. Lambert, N. Matalas, and R. Zimmerman (1999). "A study of approaches for assessing and managing the risk of extremes." *Risk Analysis,* 19(1), 83-94.

Gerzon, M. (1996). *A House Divided: Six Belief Systems Struggling for America's Soul.* New York: G.P. Putnam's Sons.

Green, T.H. (1880). *Liberal Legislation and Freedom of Contract.* Reprinted in *The Works of Thomas Hill Green* (1888), Volume 3. London: Longmans, Green and Co.

Hartz, L. 1955. *The Liberal Tradition in America.* New York: Harcourt, Brace & World, Inc.

Hobbes, T. (1651). *Leviathan.* Reprinted in *Great Political Thinkers* (1966), edited by W. Ebenstein. New York: Holt, Rinehart, and Winston.

Kaplan, S. (2003). "Applying the general theory of quantitative risk assessment (QRA) to terrorism risk." *Risk-Based Decisionmaking in Water Resources X.* American Society of Civil Engineers, Reston, VA.

McCloskey, H. and J. Zaller (1984). *The American Ethos: Public Attitudes Toward Capitalism and Democracy.* Cambridge, MA: Harvard University Press.

Mill, J.S. (1859). *On Liberty.* Reprinted in *Classics of Modern Political Theory* (1997), edited by S. Cahn. New York: Oxford University Press.

NRC (2002). *Terrorism: Perspectives from the Behavioral and Social Sciences.* N. Smelser and F. Mitchell, (Eds.). National Research Council, Panel on Behavioral, Social, and Institutional Issues, Committee on Sciences and Technology for Countering Terrorism. Washington, DC: National Academies Press.

Smith, A. (1777). *An Inquiry into the Nature and Causes of the Wealth of Nations,* The Modern Library edition (1937). New York: Random House.

GIS Model for Estimating Dam Failure Life Loss

Maged Aboelata, Student Member ASCE,[1] David S. Bowles, Fellow ASCE,[2] and Duane M. McClelland, Member ASCE[3]

Abstract

This paper describes a modular geographical information system (GIS) model that is being developed for estimating potential loss of life from natural and dam-failure floods. The model can be used to provide life-loss estimates for use in dam safety risk assessments, including dam failure caused by terrorism. It can be used to explore options for reducing life loss by reducing the probability of dam failure or by improving the effectiveness of emergency planning and response. The modeling system comprises the following internal modules: 1) loss of shelter, including prediction of structural performance of buildings, 2) warning and evacuation, and 3) loss of life based on empirical relationships developed from a wide range of case histories.

The empirical life-loss relationships have been developed for sub-populations at risk (subPar) that approximate homogenous base units (HBUs) defined in our earlier work [McClelland and Bowles 2000] and summarized in this paper. HBUs are defined by location, including different levels in buildings, type of population at risk (e.g. motorists, pedestrians, campers, and those in buildings), evacuation factors, and flooding conditions.

Estimated inundation conditions are obtained from an external flood routing model, which may include dam break and reservoir operation capabilities. Other inputs include a digital elevation model, road layout, and information on populations at risk and on structures from census data and other sources. The user must provide information on the type of population at risk, the initial spatial distribution of each

---

[1] Research Assistant, Utah Water Research Laboratory, Utah State University, Logan, Utah 84322-8200; 435-797-3198; maboelata@cc.usu.edu.
[2] Professor and Director, Institute for Dam Safety Risk Management, Utah Water Research Laboratory, Utah State University, Logan, Utah 84322-8200, and Principal, RAC Engineers & Economists; 435-797-4010; David.Bowles@usu.edu.
[3] Water Resources Engineer, GEI Consultants, Inc., 6950 South Potomac Street, Suite 200, Englewood, Colorado 80112; 303-662-0100; dmcclelland@geiconsultants.com.

sub-population at risk, building types, and warning propagation and mobilization time distributions. Default relationships are being developed for some inputs.

Application of a preliminary version of the modeling system is demonstrated by hypothetical flood-induced and sunny-day dam failures. The two cases differ in warning and evacuation characteristics with a two-stage evacuation of areas affected by spillway discharges and dam failure being considered for the flood case. Plans for further model development and testing are summarized; these include the addition of uncertainty analysis and the development of a simplified version of the model for making preliminary estimates of life loss when less detailed input is available.

Introduction

Need for life-loss estimation. To effectively reduce life-safety risks associated with dams and natural flooding, life-loss estimates are needed for the following purposes [Bowles et al. 2003]:
- To evaluate existing and residual risks against tolerable risk guidelines;
- To assess the benefits (i.e., risk reductions) associated with structural and non-structural risk reduction measures, including more effective emergency planning and evacuation;
- To estimate the cost effectiveness of life-safety risk reduction to aid in prioritizing or justifying expenditures on risk-reduction measures.

In addition, a better understanding of life-loss dynamics associated with floods is valuable for improving the development of effective emergency action plans and emergency response plans [McClelland 2002, McClelland 2003a, b, and c].

Previous approaches. Most available approaches for estimating life loss $(L)^4$ from dam failure are purely empirical using regressions on heterogeneous global population at risk (Par), and warning time (Wt). Examples include Lee et al. [1986], Brown and Graham [1988], and DeKay and McClelland [1993]. A recent approach by Graham [1999] provides life-loss ratios and ranges for a mix of Par and large subPar based on Wt, flood severity (F), and warning effectiveness (We). Assaf et al. [1998] have developed a simulation approach, which is undergoing continued development.

These approaches are reviewed in detail in McClelland and Bowles [2002]. The empirical approaches share the following limitations:
- Many factors that change with the type of dam break or natural flooding event are not separately distinguished.
- Travel times, depths, and velocities that affect the fate of people, vehicles, and structures are based on large-scale averages.
- Par is considered for the entire area of inundation or for large subPar, which does not distinguish the many attributes that are important determinants of life loss.
- Warning time is considered as a single variable without taking into account the chain of events that must occur before a message can be disseminated, the rate of

---

[4] A list of symbols used in this paper is contained in the Appendix.

warning propagation, the extent to which the warning penetrates a community, the efficacy of the warning message, and the rate of mobilization.
- Evacuation is not considered as a separate process, and the benefits of relocation to safer shelters of those who do not evacuate are not explicitly included.

Utah State University (USU) model. The goal of the USU research project upon which this paper is based is "*to develop a practical and improved life-loss estimation approach for use in dam-safety risk assessment and emergency planning.*" Our methodology is specifically formulated to overcome the limitations of previous approaches. The overall project comprises the following phases:
1) Case history characterizations and analyses;
2) Development, testing, and demonstration of a prototype simulation modeling system;
3) Development, testing, and demonstration of a simplified technique; and
4) Development, testing, and demonstration of software for the simulation modeling system and the simplified technique.

Work on Phase 1 is complete, although additional case histories could be characterized. A deterministic version of the simulation system has been completed and is being demonstrated for a Corps of Engineers dam. Work has commenced to develop a Monte Carlo (uncertainty) version of the simulation model. This version will propagate various sources of uncertainties in model inputs through to the life-loss estimates. It will also be used to generate a synthetic database for various representative inundation settings, which will be used to develop the simplified technique.

Outline of paper. The first major section of this paper summarizes the foundation for our life-loss estimation model; this lies in a very detailed characterization of flooding case histories and the development of an approach that overcomes the major limitations of previous empirical methods. The second major section provides an overview of the simulation modeling approach and its four modules. The remaining sections present some preliminary model results and planned model developments, followed by a summary and conclusions.

Foundations of USU Model

Case histories. The first phase of our research involved the collection and characterization of case histories of flood events and the people in those floods. To date, we have identified about 180 flood events that have caused loss of life. Most of these involved failure of a dam, but some were dike failures, flash floods, regional floods, or other types of floods. The characterization of each event entails dividing the population at risk (represented by the symbol Par)[5] into sub-populations at risk

---

[5]Population at risk (Par) quantifies the number of people who, without evacuating, would remain within those regions of the flood's imprint that exceed some minimum criteria of depth and velocity.

(which are called subPar or Par_i)[6] and their corresponding threatened populations $(Tpar_i)$[7], assigning values to nearly 100 quantitative or categorical (descriptive) variables[8] for each subPar, and documenting insights into life-loss dynamics. So far, 54 events have been characterized, yielding 253 non-overlapping subPar. The proportion of lives lost within these subPar ranged from zero percent to 100 percent, with good representation throughout this range.

The work of characterization has been accomplished in stages, allowing for an evolutionary, iterative process. Variables have been added, discarded, and more carefully defined as the work has progressed. Those variables that are most important to our life-loss model were developed after approximately half of the subPar were characterized. At that point, it became apparent that more traditional life-loss estimation variables, most of which were developed with heterogeneous Par in mind, failed to adequately describe dominant life-loss patterns on the scale of subPar. In subsequent characterizations, we have evaluated the usefulness of the new variables for estimating life loss and refined their definitions to improve their usefulness for life-loss estimation.

Key variables used in USU model. The scope of this paper does not permit us to fully present the subtleties and nuances of defining, estimating, or illustrating the variables used in our model; however, this section summarizes some important aspects. In particular, we discuss the dependency of the magnitude of dam-failure life loss on whether people successfully evacuate and whether those who fail to evacuate are able to find adequate shelter. More detailed presentations can be found in McClelland [2000], and McClelland and Bowles [2000; 2002].

Our research has shown that warning time (Wt)[9] is a relatively poor predictor of whether or not people will successfully evacuate.

---

[6] subPar (Par_i) are any subsets of Par.
[7] The threatened population (Tpar) quantifies members of Par that remain in the flood zone when flooding exceeds minimum criteria of depth and velocity. $Tpar_i$ is the threatened population within Par_i.
[8] It was never our intent to use all of these descriptive variables for life-loss estimation. The variables used for estimation are a subset of the entire group of descriptive variables. Those variables that are not used for estimation are useful for general understanding of life-loss dynamics and for appreciating the setting of case histories upon which the empirical aspects of our model are built.
[9] Warning time is defined in different ways. Life-loss researchers define warning time from the standpoint of the flood victim. A typical definition is the length of time from when the first public warning is issued until the dam-failure flood wave reaches the first person in the population at risk. From this perspective, warning time says something about the time potential flood victims have to evacuate, but warning time does not indicate whether or when individual warnings are received or believed. Emergency action planners sometimes define warning time from the standpoint of emergency response personnel: the time beginning at dam failure and ending when the first warning is issued.

A more successful measure of evacuation potential is the *excess evacuation time* (E):

$$E = Wt_{avg} - Ret$$

In which: E = *Excess evacuation time* (may be negative)
Wt_{avg} = *Average warning time* from all sources, including non-official sources such as neighbors and sensory clues
Ret = *Representative evacuation time*

Excess evacuation time is the average time *available* to evacuate minus the representative time *required* to evacuate. The excess evacuation time may vary dramatically from one location to another based on how quickly emergency management officials can deliver individual warnings; how urgent, credible, and frequent the warnings are; the nature of sights, sounds, and vibrations that provide natural warnings; selected modes of evacuation; the mobility of the population in question; the size of family groups; the distance to safety; barriers such as fences and bridges; and many other factors.

For those who fail to evacuate, survival usually depends on the ability to reach adequate shelter from the flood. Evacuation modeling provides a means of estimating the number of people who are at diverse locations while in the process of evacuation when they encounter flooding. Both empirical and analytical approaches, including more complex methods such as transportation evacuation modeling, provide means of estimating excess evacuation time and evacuation rates.

Lethality zones distinguish physical flood environments in which historical rates of life loss have distinctly differed. We have defined three lethality zones: chance zones, compromised zones, and safe zones. Each lethality zone is physically defined by the interplay between available shelter and local flood depths, velocities, and debris. Figure 1 shows the historical rates of life loss in the three lethality zones.

- In *chance zones*, flood victims are typically swept downstream or trapped underwater, and survival depends largely on chance; that is, the apparently random occurrence of floating debris that can be clung to, getting washed to shore, or otherwise finding refuge safely. The historical rate of life loss in chance zones ranges from about 50 percent to 100 percent, with an average rate over 90 percent (Figure 1).
- In *compromised zones,* the available shelter has been severely damaged by the flood, increasing the exposure of flood victims to violent floodwaters. An example might be when the front of a house is torn away, exposing the rooms inside to shoulder-high flooding with fast velocities. The historical rate of life loss in compromised zones ranges from zero to about 50 percent, with an average rate near 10 percent (Figure 1).
- *Safe zones* are typically dry, exposed to relatively quiescent floodwaters, or exposed to shallow flooding unlikely to sweep people off their feet. Depending on the nature of the flood, examples might include the second floor of residences and sheltered backwater regions. Life loss in safe zones is virtually zero (Figure 1).

Lethality zones are critical determinants of life-loss rates. Far more people die in chance zones than in any other type of zone. By contrast, with or without official warnings, people in safe zones are likely to survive. Generally speaking, the type of available shelter is more important than stream velocities, and stream velocities are more important than flood depths when assigning lethality zones, but all three components must be considered for their interaction.

There are two primary variables that help to define the available shelter: Par type (Pt) and loss of shelter (Ls). *Par type*[10] categorizes a subPar by its physical environment. Par type categories include population centers with buildings, campgrounds, recreation areas used for fishing or walking, automobiles, trains, and boats. *Loss of shelter* characterizes the extent to which damage to a building[11] exposes the occupants to the full force of the flood. Loss of shelter is a parent set of lethality zones. For example, high (total) loss of shelter produces a chance zone; medium (partial) loss of shelter will most likely produce a compromised zone, but safe zones or chance zones may exist at other locations in the building or even on the same story; and low loss of shelter produces a safe zone. Loss of shelter is a function of the durability and elevation of structures as they interact with flow depths, velocities, flow duration, and debris. Because only $Tpar_i$ experience lethality zones, lethality zones cannot be considered apart from excess evacuation time or an evacuation model.

Lethality zones as approximations of homogeneous base units. A central goal of our historical life-loss research is to identify dominant variables that have historically governed survival and life loss, and to derive empirical relationships based on those variables that are useful for life-loss estimation. A fundamental challenge in historical life-loss research is that there are relatively few populations at risk (Par) that have experienced flooding consistent with a dam failure, and those that have often differ greatly with respect to those variables that dominate life loss. The result is that the available sample size is small, and it is taken from many different populations in the statistically heterogeneous sense.

Lethality zones provide the means by which statistical sampling can be improved. By considering flood events at the level of lethality zones and the threatened subpopulation at risk that remains after evacuation ($Tpar_i$) we have transformed 54 heterogeneous flood events into approximately 250 homogeneous

---

[10]Technically, Par type (Pt) categorizes a sub-population at risk (Par_i) in a defined physical environment, but Pt may also be used to categorize the physical environment itself. The way Pt is categorized may vary. For example, with respect to warnings, Pt is primarily categorized based on the speed and reliability with which a warning can be delivered and the rate at which it can be propagated through the population. With respect to evacuation, Pt is primarily categorized based on the mobility of a population and its distance from safety. With respect to survival of those who fail to evacuate, Pt is primarily categorized based on the quality and availability of shelter (lethality zones). As defined in this paper, Pt is a broad category used as an initial classification of flooded regions.

[11]Loss of shelter can be applied by analogy to outside shelter – e.g., trees and islands.

flood contexts. The result is a substantial sample size representing each of the three types of lethality zones, with each sample drawn from a relatively homogeneous population.

When lethality zones are truly homogeneous in every respect, they are called *homogeneous base units* (HBUs). HBUs are an ideal construct that can only be approximated, but which is useful for descriptive purposes. HBUs are analogous to subatomic particles. Three types of particles (i.e. protons, neutrons, and electrons) are the basic building blocks of matter, regardless of how different objects may appear. We cannot estimate both the location and velocity of any one particle at any point in the future with certainty, but we can describe reliable patterns of behavior. Uncertainty is further reduced when we sum the behavior of individual particles across a larger element or compound, thus providing a form of averaging.

By analogy, three HBUs, approximated by chance zones, compromised zones, and safe zones, are the basic flood contexts from which life loss in flood events can be estimated, regardless of how different two flood events appear on a macro scale. We cannot estimate the future location and evacuation trajectory of individuals with certainty, but we can describe reliable patterns of behavior. Uncertainty is further reduced when life-loss estimates based on probability density functions for each HBU are summed across a population at risk, thus providing a form of averaging.

Analysts can approximate HBUs by dividing a population at risk into increasingly homogeneous subunits. Isolating Par_i by location promotes homogeneity on all levels. Distinguishing Par_i by Par type (Pt) minimizes differences in the physical environment. Reducing Par_i to $Tpar_i$ based on excess evacuation time or evacuation modeling isolates those who will be present in the HBUs at the time of the flood wave arrival. This significantly reduces differences in temporal-spatial flood dynamics prior to application of life-loss functions. At the level of $Tpar_i$, an analyst can approximate HBUs by identifying flood zones and estimating flood-zone densities (number of people per lethality zone). It is important that lethality zones are evaluated three-dimensionally, since in terms of proportional life loss, the HBU on the second or third story of a building might be the same as the HBU in shallow flooding near shore. Each lethality zone has a life-loss rate probability density function represented by those in Figure 1.

Simulation Model Overview

The life-loss estimation model is structured as a modular system around a database. The advantage of this approach is that it divides the system under investigation into components, thus making it easier to deal with each component individually rather than considering one complex system. A modular modeling system also makes it relatively easy to interchange alternative modules. Each module exchanges data with other modules through the database, which includes various geographical information system (GIS) layers and tables.

The simulation modeling system utilizes readily available GIS information on road layout and on population and structures from census data. Four major modules have been developed in the simulation modeling system, as follows:

1) *Inundation module*—interfaces with an existing inundation model to provide a set of grids representing water depth and flow velocities over the entire study area and throughout the duration of the flood event.
2) *Loss-of-shelter module*—simulates the exposure of people in buildings during each flood event as a result of structural damage, building submergence, and toppling of people in partially damaged buildings. Loss of shelter and flood zone categories are assigned to each building level[12] in various types of buildings throughout the inundation area.
3) *Warning and evacuation module*—simulates population redistribution following issuance of a warning. Redistribution includes lateral evacuation along evacuation routes from each subPar[13], and vertical relocation within the inundation area to accessible shelters. Times when vehicles and pedestrians are expected to become unstable, such as by flotation of vehicles or toppling of pedestrians, are estimated along evacuation routes. After those times, mobilization ceases, and people in vehicles or on foot along evacuation routes are assigned to a chance flood zone.
4) *Loss-of-life module*—based on the assigned flood zone categories, life-loss estimates are made using life-loss probability distributions in Figure 1.

The model is designed to be applied to a set of event-exposure scenarios. Events include different dam failure modes and locations, no-failure flooding, and different flood severities. Exposure cases can include different seasons, day/night, and weekend/weekday.

The following subsections provide some additional details on each of the four modules.

Inundation module. Inundation modeling is the first step in estimating loss of life. It is considered as an external module, because existing models, such as DAMBRK or HEC-RAS, are used. Inundation results are transferred to the GIS database in the form of a set of grids representing 1) water depth and flow velocity throughout the study area and 2) over a time period that covers the passage of the dam-break flood wave throughout the study area and 3) a no-failure flood wave in the case of flood-induced failures.

Loss-of-shelter module. Loss of shelter estimates the exposure to the flood water of the population at risk. Previous studies concerned with the performance of structures in floods, such as Black [1975], RESCDAM [2000], Sangrey [1975], and USACE [1985], have provided ways to estimate the partial or complete destruction of a building as a function of depth and velocity conditions for different types of structures with different numbers of floors. Heavy structures, such as masonry or concrete, have a higher resistance to destruction or flotation in a flood than lightly constructed wooden structures. Structures that are anchored to their foundations have a higher resistance than unanchored structures.

---

[12]Building levels can include various floor levels, the basement, and the roof.
[13]Census blocks are being used as subPar.

Flood depth and velocity determine the loss-of-shelter category in two ways. First, flood depth and velocity can cause physical damage to a structure. Second, they can lead to submergence of some or all building levels, which also causes loss of shelter to Tpar$_i$, regardless of physical damage to the structure. Submergence is defined as a water level inside the structure that makes survival very unlikely. Unlike structural damage, which is an attribute of an entire building, submergence is defined separately for each building level. Figure 2 summarizes the process of assigning loss-of-shelter categories to building levels of each type of structure at successive time steps throughout the flood event.

Warning and evacuation module. One of the most important factors in estimating expected life loss is the location of the threatened population at the time when the flood arrives. When time permits, people will usually attempt to laterally evacuate the area before flood arrival in response to official or unofficial warnings. Evacuation effectiveness depends on many factors discussed above. Evacuation can also be by vertical relocation within a structure.

Figure 3 is a schematic of a typical warning and evacuation process represented in the model. It comprises various time lines for the following events associated with three entities which are involved in a dam failure and inundation scenario:

- The *dam* for the following events: detection of a failure or potential failure; decision to notify the authorities in each emergency management area (EMA); notification; and dam failure.[14]
- Each *emergency management area* for the following events: receiving a notification from the dam owner; decision to warn the public; and issuance of a warning.
- Each *subPar (Par$_i$) within each EMA* for the following events: receiving a warning; mobilization; and clearance of the flood inundation area. A stage hydrograph is shown in Figure 4 for the first subPar (Par$_{A1}$) to represent the arrival of dam-break flows at various locations in the subPar as the flood wave moves downstream.

The process starts at the dam with the development of a failure mode or a prediction that dam failure is likely to occur. The prediction could be the result of an observed condition, such as excessive seepage containing fines, or a forecast condition, such as a major inflow flood that is projected to overtop the dam. The owner is responsible for detecting the failure mode at the dam and notifying the responsible authorities. However, in some cases a failure mode may be detected by someone other than the owner or operator's representative. In such cases the failure mode, whether an incipient condition or a failure that has already occurred, may be reported directly to the authorities by the person who observes unusual conditions at the dam or a dam-break flood wave downstream. Figure 3 shows the case of failure taking place after

---

[14]The order of these events may vary. For example, detection may not take place before failure.

notification, although in other cases it can occur before notification or even before detection.

The notification of a dam failure or expected failure is forwarded to emergency managers along the flood path. These managers are responsible for issuing warnings to the various subPar within their jurisdiction. People who receive the warning, and who are willing to evacuate, will mobilize, choosing their mode of evacuation. Our model provides for evacuation by vehicle or on foot. The communication of informal warnings is commonly reported in historical flood events. It is planned to add into the model a capability to represent this phenomenon.

Alternatively, people may decide to go upstairs in the same or a nearby building, although the model does not explicitly represent people moving between buildings. Moving upstairs is referred to as "vertical relocation" or "vertical evacuation" to distinguish it from lateral evacuation out of the inundation area. The success of vertical evacuation will depend on how the building performs under the imposed flood loading. The flood wave may overtake some people who are evacuating. The model categorizes people who are in vehicles or on foot as being in a chance flood zone if flooding conditions are estimated to be unstable for vehicles or pedestrians (Figure 4). Beyond the time at which vehicles or pedestrians become unstable, no further lateral evacuation is represented by the model. Thus, this module redistributes the population at risk from its initial distribution by Par type at the time that a warning is issued, to new distributions with assigned flood-zone categories.

Loss-of-life module. Loss of life is estimated by applying the probability of exceedance relationships for percentage of fatalities in each flood-zone category (Figure 1) to the $Tpar_i$ in each census block estimated using the warning and evacuation module. The spatial and temporal distributions of estimated life loss for an event-exposure scenario can be shown based on census block estimates for structures, and evacuation pathways for vehicles and pedestrians. Numerical estimates can be shown in aggregate form for the entire inundation area or broken out by communities or other sub-areas. At present, estimates are based on the average fatality rates for each flood-zone category. When the Monte Carlo version of the model is completed, estimates will be presented as probability distributions of life loss.

An Example

As a preliminary illustration of the life-loss estimates from the deterministic version of the model, we have included some results for a community of about 3,500 for hypothetical dam breaks under flood and sunny-day conditions. The community is located between 5 and 8 miles downstream of the dam. Warnings are assumed to be issued 0.5 hours before the sunny-day failure and three hours before the flood failure. A Stage 1 evacuation of the flood plain affected by spillway discharges is assumed to be 100% effective in this example. Thus life loss is estimated only for the incremental Stage 2 evacuation area with a population of about 1,400 affected by flood-induced dam failure.

Figures 5 and 6 show the spatial distribution of life loss by census block expressed as a number of fatalities in the upper figures and as a fatality rate in the lower figures, for the hypothetical flood-induced and sunny-day failures, respectively. Figures 7 and 8 provide a tracking of the overall population at risk from its initial location in structures[15], through the warning and evacuation processes, through to its final disposition (i.e., cleared inundation area, survived or lost life during evacuation, or survived or lost life in structures) for the hypothetical flood-induced and sunny-day failures, respectively. The effect of a staged evacuation and the longer warning time for the flood failure can clearly be seen to lead to a much smaller life loss than for the sunny-day failure.

Planned Model Developments

Uncertainty mode. In life-loss estimation there are many factors that cannot be predicted with certainty, including the responses of individuals to warnings. Therefore, we are developing a Monte Carlo (uncertainty) version of the life-loss estimation model to account for the effects of uncertainties in various model inputs, including the following: structure type distribution, structure damage criteria, structure submergence criteria, human and vehicle stability criteria, time of initial warning, warning diffusion/effectiveness, mobilization time distribution, mode of evacuation, initial population distribution, and fatality probability rate by flood-zone category.

The Monte Carlo version will propagate input uncertainties through to uncertainties on life-loss estimates, which will be presented as probability distributions. These probabilistic estimates of life loss can be used in dam safety risk assessments, such that the estimation errors associated with life loss and other risk assessment inputs are represented in risk assessment results, including evaluations against tolerable risk guidelines as illustrated by Chauhan and Bowles [2001]. Alternatively, expected values of estimated life loss and confidence bounds on these estimates could be used in risk assessments.

Simplified mode. The uncertainty mode of our model will be used to generate a synthetic database for various representative inundation settings. This database will be used to develop a simplified "empirical" technique for life-loss estimation. For inundation settings that are adequately represented, the simplified technique will provide life-loss estimates for a lower level of effort than needed for the full modeling system, although with greater uncertainties than the full modeling system. However, provided that GIS data sources are readily available, the level of effort needed for implementing the full modeling system is not considered to be unreasonable.

---

[15]Other initial distributions that include locations outside of structures are being added as a capability in the model.

Summary and Conclusions

Dam safety risk assessment requires credible life-loss estimates. The first phase of our research has yielded many useful insights into life-loss dynamics for dam failure and natural floods. Based in part on these findings, our paper summarizes a distributed modeling approach to consider evacuation, detailed flood dynamics, loss of shelter, and historically-based life loss. The approach uses a modular modeling system that will allow the use of different inundation and evacuation-transportation models. Some preliminary results are presented to illustrate the type of information that can be obtained from the deterministic version of the modeling system and the effects of improving warning time.

An uncertainty version of the modeling system is under development so that the uncertainties that are intrinsic to life-loss estimates can be represented in model outputs and used in subsequent risk assessments. A simplified version of the life-loss estimation procedure is to be developed using synthetic samples from the uncertainty version. Using the simplified version, preliminary estimates will be obtainable with less effort in those settings that are to be covered by this technique. In addition to dam safety risk assessment applications, estimates of life loss from our modeling system can be expected to provide a basis for better understanding the effectiveness of emergency planning and evacuation in flood plains and below dams, and to lead to their improvement.

Acknowledgements

The authors acknowledge the sponsorship of the US Army Corps of Engineers (USACE), the Australian National Committee on Large Dams (ANCOLD) and many of its organizational members, the US Bureau of Reclamation (USBR), and Utah State University (USU). We also acknowledge valuable insights provided by Mr. Wayne Graham, USBR, and the provision of case histories by Mr. Graham and the Center on the Performance of Dams, National Performance of Dams Program, Stanford University. DAMBRK runs used in our example were completed by Mr. Zhengang Wang and Dr. Sanjay S. Chauhan, Utah State University.

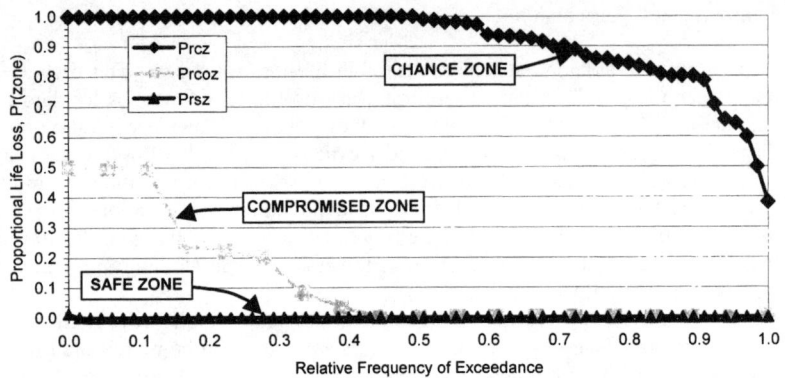

Figure 1. Historical life-loss rates in lethality zones.

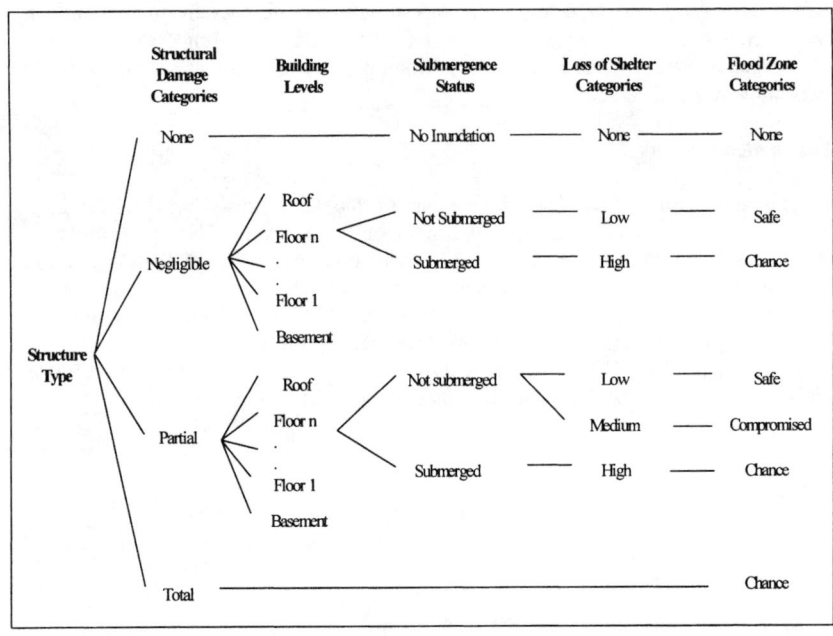

Figure 2. Assignment of flood-zone categories for building levels.

RISK-BASED DECISIONMAKING IN WATER RESOURCES X 139

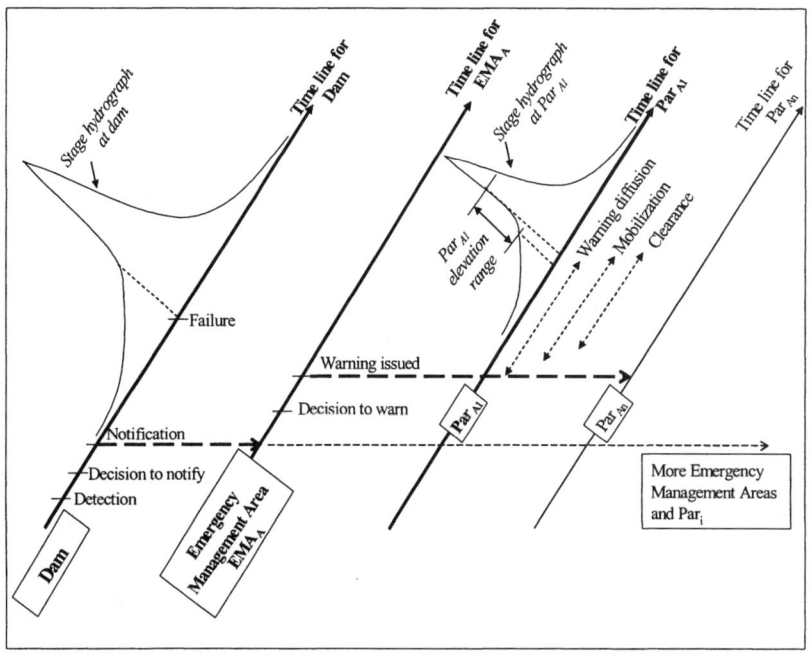

Figure 3. Time lines for events in warning and evacuation processes.

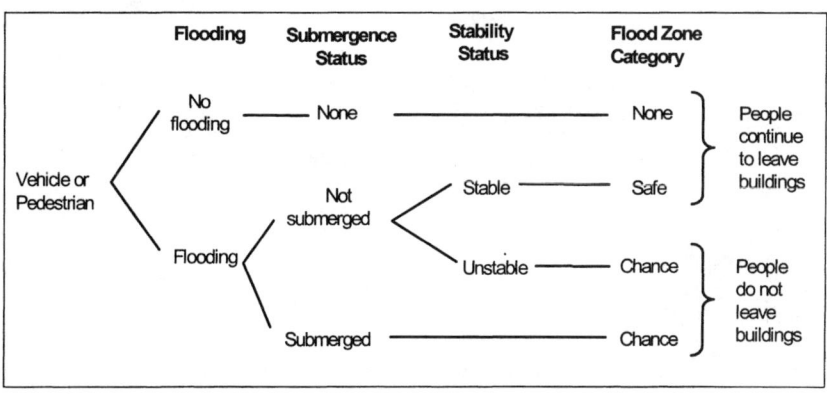

Figure 4. Assignment of flood zones for vehicles and pedestrians.

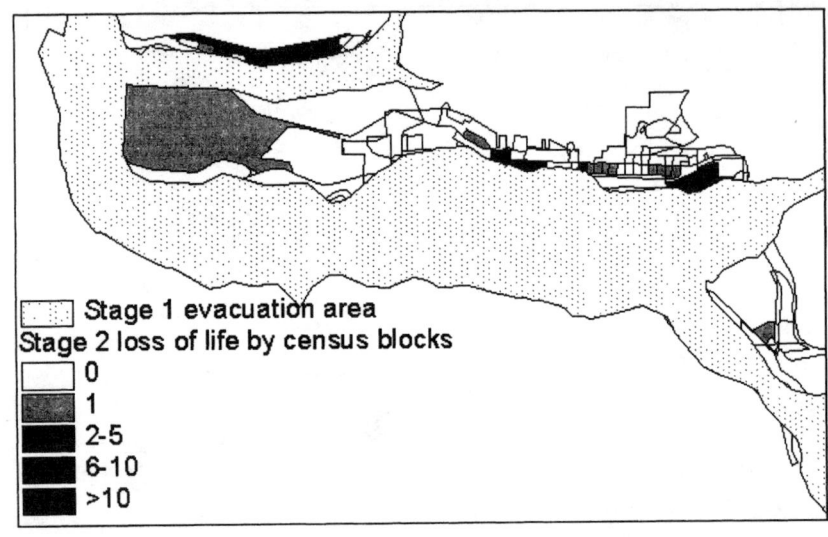
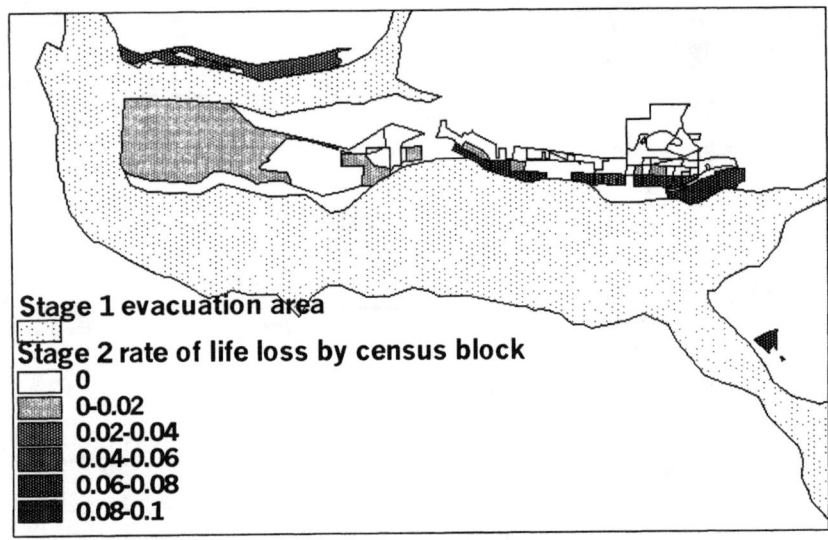

Figure 5. Example life-loss estimates for flood failure.

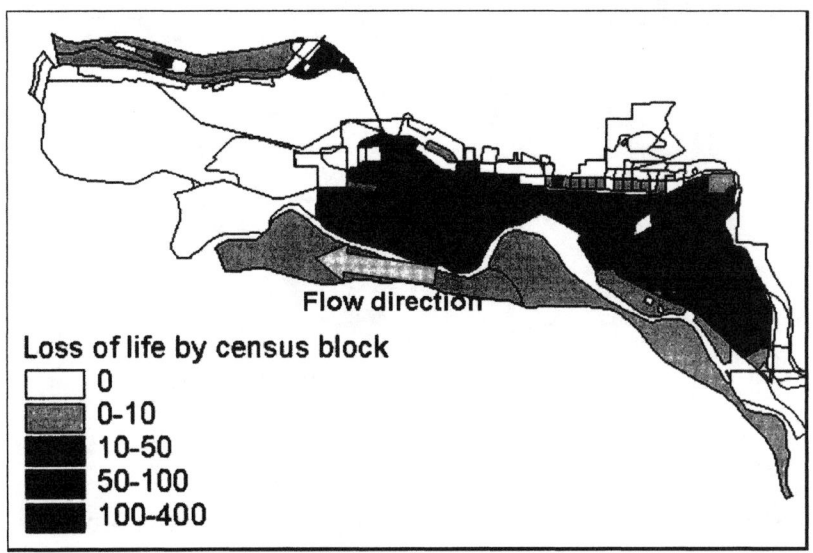

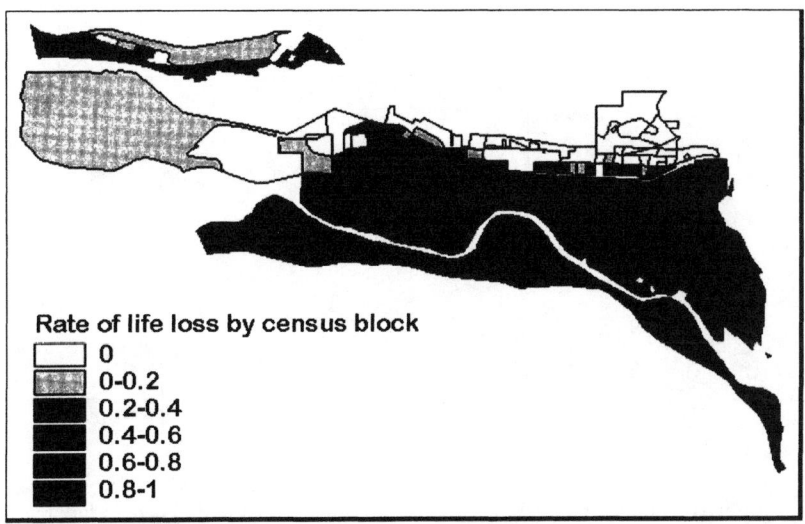

Figure 6. Example life-loss estimates, sunny-day failure.

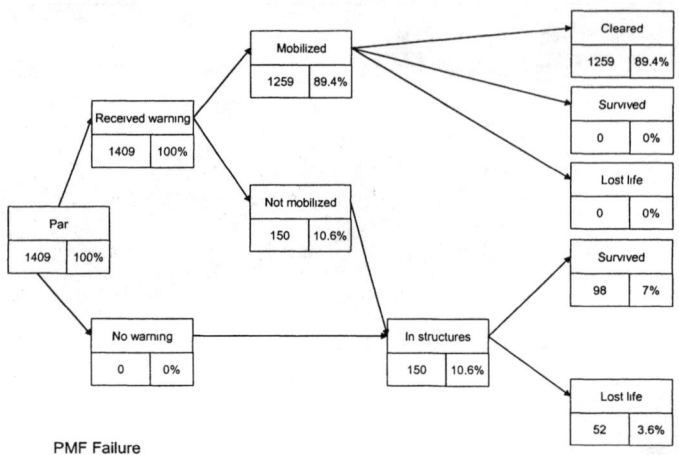

Figure 7. Schematic tracking PAR for a hypothetical flood-induced dam failure.

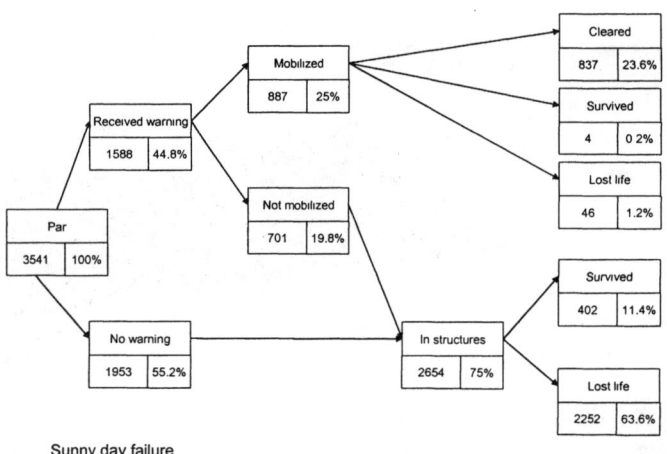

Figure 8. Schematic tracking PAR for a hypothetical sunny-day dam failure.

Appendix: List of Symbols

$\square_i$	subscript i generally indicates that the variable pertains to Par_i
Coz	compromised zone
Cz	chance zone
E	excess evacuation time (minutes)
F	flood severity
HBU	homogeneous base unit
L	life loss
Ls	loss of shelter (L = low, M = major, H = high)
Par	population at risk
Par_i	subpopulation at risk; same as subPar
Pr(zone)	proportion of lives lost in a designated flood zone
Prcoz	proportion of lives lost in the compromised zone
Prcz	proportion of lives lost in the chance zone
Prsz	proportion of lives lost in the safe zone
Pt	Par (population at risk) type
Ret	representative evacuation time (minutes)
subPar	same as Par_i
Sz	safe zone
Tpar	threatened population at risk
$Tpar_i$	threatened subpopulation at risk
We	warning effectiveness
Wt	warning time (the first formal warning, in minutes)
Wt_i	individual warning time (also, Wt specific to Par_i)
Wt_{avg}	average warning time (from any source, in minutes)

References

Assaf, H., D.N.D. Hartford, and J.D. Cattanach (1998). "BC Hydro method: estimating dam breach flood survival probabilities." In *A Guide to Risk Analysis of Dam Safety,* draft commentary 4-17–4-35. Canadian Electricity Association, Ottawa, Ontario, Canada.

Black, R.D. (1975). *Flood-Proofing Rural Residence, A Project Agnes Report, Pennsylvania.* New York State College of Agriculture and Life Sciences, Ithaca, NY. Prepared for Office of Technical Assistance, Economic Development Administration, Washington, DC.

Bowles, D.S., L.R Anderson, T.F. Glover, and S.S. Chauhan (2003). "Dam safety decision-making: combining engineering assessments with risk information." *Proceedings of the 2003 US Society on Dams Annual Lecture,* Charleston, South Carolina, April.

Brown, C.A. and W.J. Graham (1988). "Assessing the threat to life from dam failure." *Water Resources Bulletin,* 24,1303-1309.

Chauhan, S.S and D.S. Bowles (2001). "Incorporating uncertainty into dam safety risk assessment." *Proceedings of Risk Analysis in Dam Safety,* Third International Conference on Dam Safety Evaluation, Goa, India, December.

DeKay, M.L. and G.H. McClelland (1993). "Predicting loss of life in cases of dam failure and flash flood." *Risk Analysis* 13(2),193-205.

Graham, W.J. (1999). *A Procedure for Estimating Loss of Life Caused by Dam Failure.* Report No. DSO-99-06, Dam Safety Office, US Bureau of Reclamation, Denver, CO.

Lee, R., P.S. Hu, D.M. Neal, M.R. Ogles, J.H. Sorensen, and D.A. Trumble (1986). *Predicting Loss of Life From Floods,* Oak Ridge National Laboratory Draft Report. Prepared for the Institute for Water Resources, US Army Corps of Engineers, Alexandria, VA.

McClelland, D.M. (2000). *Estimating Life Loss for Dam Safety Risk Assessment.* Utah State University, Logan, UT.

McClelland, D.M. (2002). *A Team Approach to Improving Dam-Safety Emergency Action Plans, Flood Maps, and Emergency Response Plans Using Life-Loss Risk Assessment.* EAP 2002: International Workshop for Emergency Preparedness at Dams, sponsored by FERC and ASDSO, April 16-18, 2002.

McClelland, D.M. (2003a). "Optimizing EAPs and ERPs, part 1: risk-based planning." *Proceedings of the 2003 US Society on Dams Annual Lecture,* Charleston, SC. US Society on Dams, Denver, CO.

McClelland, D.M. (2003b). "Optimizing EAPs and ERPs, part 2: team planning." *Proceedings of the 2003 US Society on Dams Annual Lecture,* Charleston, SC, April. US Society on Dams, Denver, CO.

McClelland, D.M (2003c). Optimizing EAPs and ERPs, part 3: prescriptive flood maps. *Proceedings of the 2003 US Society on Dams Annual Lecture,* Charleston, SC. US Society on Dams, Denver, CO.

McClelland, D.M. and D.S. Bowles (2000). "Estimating life loss for dam safety and risk assessment: lessons from case histories." *Proceedings of the 2000 Annual USCOLD Conference,* US Society on Dams (formerly US Committee on Large Dams), Denver, CO.

McClelland, D.M. and D.S. Bowles (2002). *Estimating Life Loss for Dam Safety Risk Assessment—a Review and New Approach.* Institute for Water Resources, US Army Corps of Engineers, Alexandria, VA.

RESCDAM (2000). *The Use of Physical Models in Dam-Break Flood Analysis.* Rescue Actions Based on Dam-Break Flood Analysis, final report of Helsinki University of Technology, Helsinki, Finland.

Sangrey, D.A., P.J. Murphy, and J.L. Nieber (1975). *Evaluating the Impact of Structurally Interrupted Flood Plain Flows.* Cornell University, prepared for Office of Water Research and Technology. Report PB-247 552, distributed by NTIS, National Technical Information Service, Springfield, VA.

USACE (1985). *Business Depth Damage Analysis Procedure.* Research Report 85-R-5, Engineering Institute for Water Resources, US Army Corps of Engineers, Alexandria, VA.

Disintegrated Water Resources Management in the US: The Union of Sysiphus and Pandora

Eugene Z. Stakhiv[1]

Abstract

It is important to think of the future in a more comprehensive way. The present high-level water policies have good intentions, but little follow-through, giving them a Sysiphean aspect. We need a broad guiding national philosophy directly related to the principles of integrated water management (IWRM) and sustainable development (SD). Both IWRM and SD require a uniform set of planning and management principles, consistently applied. A *national water policy* would be a good starting-point. In the absence of a central federal water infrastructure agency or an interagency council, a *President's Water Council* may be the needed interim solution. There is a Pandora's box of current difficulties in federal water management, but we can always have hope that among the changes that are surely coming we will have positive solutions.

Introduction

"With the turn of a switch on New Year's Day, the federal government finally did what it has threatened to do for years: it reduced the flow of water from the Colorado River that has helped California coastal cities bloom." (*Washington Post,* January 6, 2002). The Clinton administration had pressed California for years to reduce its share of Colorado water and California pledged to come up with a plan by New Year's Day, 2003. But an agreement brokered by the state broke down at the last minute because of concerns of farmers in the Imperial Valley about potential liability for environmental damages in the Salton Sea, which is currently sustained almost entirely by irrigation return flows. Utilities and their customers will have to find new ways to conserve, and the cost of water will inevitably go up.

---

[1] Institute for Water Resources, US Army Corps of Engineers, Alexandria, VA 22315; (703) 428-8077; eugene.z.stakhiv@wrc01.usace.army.mil.

This scene is being replayed many times over throughout the US in large systems (e.g., Missouri River, Apalachicola River, Florida Everglades) as in small ones. Many small and seemingly disconnected decisions are being made about water management. The Colorado water reduction was a relatively harsh step for the federal government to take, but in the absence of a clear national policy on water resources, it was a wakeup call to the state of California that they must do more and take more responsibility for effective and efficient water management within their state and river-basin boundaries. This action *de facto* constitutes a policy in the sense of Arthur Maass's [1951] dictum that "[P]ublic policy is being formed as it is being executed, and it is being executed as it is being formed." Policies should be statements of societal goals, or directives for achieving goals, but in many cases today, policies are simple administrative substitutes for marginal tinkering with the status quo. Since there has been no real water "crisis" in the US as yet, this incremental adaptation may be all that is required for the time being. However, we must begin to think of the future in a more comprehensive way.

The Maass dictum is at the core of "adaptive management." It is the mantra of the "new age" water and environmental managers: learn, refine, improve, and implement as you go. It is a more purposeful form of "autonomous adaptation" that has successfully brought the following present conditions to the US: freshwater withdrawals and water use has steadily declined since its peak in 1975, streams are cleaner, and water is progressively being reallocated for ecological purposes. However, adaptive management is inherently incremental in nature, continuously spawning a plethora of new rules, good ideas (as well as bad ones), and procedures that are dispersed indiscriminately, continuously adding new threads to the growing "Gordian knot" of public policy. IWRM and SD require a future-oriented management plan—these are anticipatory and imply managing for the long term [Loucks et al. 2000]. If we seriously are to fulfill the implicit requirements of IWRM and SD, then we need to substantially rethink and rework the principles, rules, and procedures that we operate under today. A "paradigm shift" is in order, together with major institutional and organizational changes.

The Case for Reform

Before we can manage water effectively and efficiently, we must have a vision—i.e., a set of clearly stated goals that directs society towards achieving its concept of SD. Water-related services simply support the attainment of a higher-order set of societal goals that include environmental quality, economic productivity, social well-being, and equity. Whether one agrees or not with the vision in the President's Council on Sustainable Development [PCSD 1996], it presents a useful starting-point in the much-needed debate on how IWRM fits in with SD, particularly the aspects of governance which are central to IWRM. US agencies cannot readily implement IWRM because unlike the World Bank or USAID, they lack the ability to propose, much less implement top-down institutional changes as part of their water resources planning and design activities. Water resource solutions directed by federal agencies are essentially "patches" on an outdated plumbing system.

Federal water agencies such as the US Army Corps of Engineers (Corps), e.g., are expected simply to build projects that provide additional capacity, to either supply more water, increase or reallocate storage, enlarge navigation locks, or offer more hydropower or greater flood protection. Providing water-supply storage to a municipality without requiring additional efficiencies such as water conservation measures, drought contingency planning, interconnections with other systems, or water pricing will not lead to sustainable development. If neither the Corps, Bureau of Reclamation, nor the National Resources Conservation Service (NRCS) are allowed to recommend institutional changes as part of their project planning and water management responsibilities, then another entity must be created to fill the void. River basin commissions or watershed councils can be empowered to do so. These would be inherently political in nature, representing federal, state, and local needs, with the authority to propose legislation for needed water reallocation and other necessary institutional changes. Under this institutional framework, the existing (or reorganized) federal agencies would simply implement the infrastructure components of a given watershed management plan, either through grant programs for small projects or construction and management of larger-scale, interstate projects. Of course, a *national water policy* and *national water code* would be needed to guide the watershed councils as well as the implementing agencies.

On its own, no single agency can initiate major water policy reform through its legislative process; for this reason, much-needed institutional changes are fought over indirectly through the project-planning process. This results in major inefficiencies because the implementing agencies do not have the legislative authority to accommodate those changes. As a consequence, large projects have many add-on features that are there simply as political compromises to satisfy numerous interests with veto power. Reform has to come from the outside, usually from a commission, but the time must be right for such reform, coupled with a willingness to undertake the agony of numerous political battles over each new idea. There have been many commissions during the past 25 years, but very little reform by Congress. Congress prefers to tinker with the margins. There is the National Energy Efficiency Act of 1992, which mandates 1.6-gallon flush toilets, but *no national energy policy*. Each biennial Water Resources Development Act specifies the depths of each navigation channel or the storage acreage allocated to municipal water supply, but *no national water policy*. The organization of the new Department of Homeland Security, which combined the functions of 100 federal organizations in 22 departments into a bureaucracy of nearly 170,000 civil servants, may have opened the door to other such reorganizations. A comparable consolidation of water programs and functions alone may suffice as the needed first step in improving the effectiveness and efficiency of water resources management, at least from the federal perspective.

Regardless of how all of these changes play out over the next few years, it is much more helpful for federal agencies, states, and localities to have a clear understanding of federal water policy goals, objectives, and criteria, *a priori*, rather than have them evolve incrementally, decision by decision. At a minimum, there is a need to reconcile the principles and weed out the conflicting rules that have accumulated in thousands of pages of Federal Register rules and regulations, reflecting the inconsistencies of numerous pieces of legislation and selective agency interpretations of that legislation. A *national water code* is needed to consolidate and simplify the labyrinth of arcane rules and procedures. If water managers are to implement the principles of IWRM and SD, they deserve to have the basic principles and policies of water management clearly presented, at least from the federal level.

Our nation cannot continue to muddle through in a disjointed incremental manner [Lindblom 1959], relying on lots of money and time and infinite patience to compensate for, mitigate, ameliorate, and politically paper over whatever problems we encounter. The hugely complex and outrageously expensive $8 billion Florida Everglades Restoration Plan is a classic example of "buying" political support for a needlessly complicated and inefficient plan. Reducing subsidies to the sugar-growing industries in that region probably would have a greater positive environmental impact than the collection of disparate restoration measures, and at much lower cost to the nation's taxpayers. Which institution or agency is responsible for raising this issue as part of the restoration planning process? How are all the non-structural, institutional, and consumption issues which have a direct bearing on resource uses in the Everglades region dealt with in the context of a more straightforward "plumbing problem?" Why should the taxpayers in Wyoming subsidize bad planning, poor analysis, and "boutique" ecological design in Florida? Will this US administration deal as firmly with Florida as it did with California? There are signs that the proposed plans for Everglades eco-restoration will finally be subjected to more rigorous scientific and economic cost-effectiveness analysis, where little had been done before. This is just in time, for there are several other large-scale, multi-billion-dollar eco-restoration efforts on the drawing boards in several other states.

The federal government is responsible for establishing the "rules of the game," since it maintains much of the basic water resources infrastructure and many of the present rules exist as federal regulatory constraints on resource uses. Our recent neglect of federal water policy and our focus on project and site-specific problem-solving is contrary to integrated water management. The most basic goal of SD and IWRM is the responsibility to live within one's means and use resources efficiently. That's what management is meant to do, and IWRM is the management basis for achieving sustainable development. Does the federal government have to come in and bail out every community or state which has failed to implement the simplest management measures? Isn't the principle of *subsidiarity* a fundamental mainstay of natural resources management (and of *federalism* itself)? Subsidiarity and federalism both hold that decisions and responsibilities for management should be made at the levels closest and most appropriate to the users of the resources. That is precisely the point of having watershed councils and local representation in numerous conservation districts, irrigation and flood control districts, and a host of other locally-oriented management mechanisms.

In the past decade, since the Long's Peak [NRLC 1992] water forum, there have been several important national water policy forums, numerous lesser efforts by various think tanks, and National Academy of Sciences studies addressing the basis for water management reforms. The Galloway Report [IFMRC 1994] on the 1993 flooding of the upper Mississippi River had great promise, with over 800 specific policy and technical recommendations for flood damage reduction. A few of these were implemented by the respective agencies, but many more substantive recommendations were ignored. The Western Water Policy Review Advisory Commission came out with a report [WWPRAC 1998] reflecting an accumulation of water management policy proposals since the National Water Commission [1973] and President Carter's Water Policy Initiative [1977]. That was followed by the report of the National Drought Policy Commission [NDPC 2000]. Also, there have been an ongoing series of National Research Council reviews of Corps of Engineers' water resources planning and management practices with an eye towards "reform." These would be useful inquiries if the Corps were in the dominant position that it enjoyed in the 1950s.

It is not the Corps that needs reform so much as federal water resources management, which includes all of the impinging regulatory programs that affect the much larger private sector demands on water and natural resources. Neither the Corps nor the Bureau operate as independent agencies—their actions, by necessity, reflect contemporary conflicting policies, values, and mandates of the administration and Congress, and they are under the tight budgetary control of the Office of Management and Budget (OMB). The water management system in the US is simply too fragmented to even begin to understand where and which "reforms" might have the greatest payoff. Increasing non-federal cost-sharing and tightening project justification requirements have been the two most effective and routinely applied management instruments that every administration has used to control an agency's programs. For example, in year 2000 dollars, the Corps' program declined by 50% from its peak in 1978, with nearly 50% of the budget going towards operation and maintenance of existing facilities and nearly 20% dedicated to environmental restoration. This last is anticipated to grow dramatically with the inclusion of several very large eco-restoration projects that are in the planning stages. However, unlike with the nation's economy, the desired outcomes of water resources management cannot be achieved simply through the equivalent of manipulating the federal discount rate.

Unfortunately, since the Water Resources Development Act of 1986, little has been done to address these important institutional policy issues by either the administration or Congress, each being content with a piecemeal "pork-barrel politics" approach to problem-solving. This is the Sysiphean aspect of policy-making: a succession of high-level water policy forums with good intentions, but little follow-through. It is ironic that both the US Agency for International Development (USAID) and the World Bank have more success in advocating IWRM in developing nations then we have in the US. For both agencies, the starting point for IWRM is a *national water policy*. It is ironic that when foreign water ministry officials visit the US to learn about our system, it is very difficult to explain the fragmentation and complexity of our water management system: the multiple layers, numerous congressional committees, countless programs, and complex cost-sharing rules. What examples of good IWRM and what advice can we possibly give to over 100 developing nations after they've seen the Florida Everglades restoration effort? It would be interesting to invite either the World Bank or USAID to undertake the water management planning effort for one river basin. A good starting-point would be simply to use the World Bank's water policy document [World Bank 1993] as a template to determine whether the proposed Everglades restoration meets their principles for integrated water management.

What is the likelihood that anything will be done in the next few years to move toward water policy clarification? Unfortunately very little, and in fact, we may see a move in the opposite direction. There is a confluence of events that is equivalent to the opening of Pandora's box; in the near term (5-10 years), the consequences of this are likely to push water policy reforms further into the background. The contemporary Pandora-policy "plagues and furies" are:
1. President Bush's management initiatives (similar to Vice President Gore's performance initiatives),
2. the resources required and attention paid to coping with global problems in foreign policy and the war on terrorism,
3. homeland security and its implications for reorganizing large bureaucracies,
4. impending budget deficits and what this implies for discretionary programs such as water resources and natural resources management, and
5. Department of Defense (DOD) privatization initiatives and consequences for the Corps of Engineers.

Each agency is currently having a difficult time simply incorporating the present administration's rhetoric into their "strategic plans," which were just completed under the previous (Clinton) administration. While the Bush administration may not be able (and certainly not the Corps or Bureau) to line up all the policy-related "ducks" in order and tackle water resources management in a comprehensive and integrated manner, the hard-line approach towards California may auger an implementation strategy catalyzing the necessary prerequisites for systemic reform. It would set the stage for policymakers to rethink their views of immediate "customer satisfaction" in favor of the longer-term view of the "public interest". Over the past decade, we've moved from being "citizens" to "customers" [Crenson and Ginsberg 2002]. Increasingly, public officials view the electorate as customers who receive public services rather than citizens who own the government. This business-oriented model has its place and value, but it results in people and communities seeking the best deal. This serves to subvert one of the basic purposes of government, i.e., equity and fairness. If a community or an interest group can go "agency-shopping" for the best deal, exploiting the differences in implementation rules and regulations for the same resource (water, wetlands, hydropower, etc.), then that factor alone undermines the achievement of SD and IWRM.

The customer satisfaction mindset is everywhere, in politics and in agency mission statements. Vice President Gore's report [NPR 1993] on reinventing government highlighted customer satisfaction, and the Corps' former Chief of Engineers echoed that refrain in his "vision" statement by declaring that customer satisfaction was not enough for him—his goal was that the customer be "delighted." Delighting individual customers at the expense of the national or public interest is not the mandate of a federal agency. Project cost-sharing, whether it is 50-50 or 70-30, still requires the federal participants in any water project to represent the national public interest. SD and IWRM require a uniform set of planning and management principles that are consistently applied, whether in California or Florida. That is not the case today, as decisionmaking is very much site-specific and skewed to the interests of the local sponsors.

For the past 20 years, both professional water managers and academicians have felt the loss of the US Water Resources Council. It was a policy forum and served as the apex of an evolving watershed-based hierarchical management framework that was thought to be an essential component of IWRM. During this time, numerous academicians and commissions have promoted the idea of reinstituting this essential organization as a way of coordinating the growing number of programs, conflicts, and policies. Instead, successive administrations have exhorted agencies towards greater interagency collaboration, cooperation, and coordination as a substitute for true reform. The approach inherently recognized that substantive and comprehensive water management reform was too difficult to undertake. If we are interested in improving the status quo, recognizing the hard realities and immense difficulties of institutional reform, then a *President's Water Council* (in the absence of a central federal water infrastructure agency or an interagency water resources council) may be the needed interim solution to the current difficulties in federal water management. In his book, *America's Water: Federal Roles and Responsibilities,* Rogers [1993] stated, "The primary responsibility of such a council would be to formulate a coordinated approach to water resources nationally and to oversee the federal contribution to that effort." This council at least would be in the position of advising the president on the best strategy for reorganizing federal water and natural resources management.

The interim President's Water Council (PWC), supported by technical staff, would be equivalent to the Council of Economic Advisors (CEA) and the Council on Environmental Quality (CEQ), the chairs of which also should be members of the PWC. Other members could include former senior-level officials of the World Bank and USAID, as well as former governors and senators. Their goal would be to tackle the reorganization of agencies and functions, separating operational and construction functions from regulatory ones. They would also propose the "organic" legislation necessary to complement the institutional changes in order to fully implement IWRM, including river-basin commissions and watershed councils. Such legislation would include the development of a national water code and a national water policy, new project-evaluation methods, determining who collects and analyzes water data, who conducts periodic national water assessments and river-basin assessments, and who forecasts stream flows and climate-change impact assessments. The first task would be to scour through all the recent commission reports, National Academy studies, and GAO reports to cull out all the useful ideas that have been developed over time.

Conclusion

Changes are coming, many of which will indirectly affect water management in the US over the next decade and beyond. Many will be positive, while others are likely to be adverse unintended consequences of policy decisions made for the right reasons. However, the temptation to cut the "Gordian (policy) knot" is very strong today. Wholesale privatization of planning, policy, and regulatory functions that are inherently governmental, as proposed by DOD Secretary Rumsfeld and the OMB, is simply the wrong approach to take for water resources management. On the other hand, consolidation of water management functions needs to be seriously considered, as proposed in the very recent report of the Volcker Commission [NCPS 2003], as well as in earlier works by Rogers [1993] and others. Can any administration hope to undertake successfully the broad sweeping changes of all federal functions simultaneously? Most definitely not. Simply implementing the Department of Homeland Security will cause major headaches for Congress and the administration. Organizational changes in water and environmental resources will have to wait in line behind the more pressing and difficult issues of tax reform, health care and coverage, social security and education reforms, and prosecution of the various foreign policy and military conflicts.

After all the evils that beset mankind had flown out of Pandora's box, there was one attribute that remained behind—*hope*. Hopefully, the next administration can begin to tackle the natural resource management issues that confront this nation. For that it needs a coherent plan. The opening moves of this administration's "Gordian knot gambit" are beginning to reveal an incremental strategy of improvements in internal agency performance and accountability. The government would do much better if it recognized that over the next two years, a President's Water Council could organize a coherent approach to a new national water management strategy, setting the stage for legislative action and reorganization in the next administration.

Note: *this paper is not intended to represent the policies of the US Army Corps of Engineers or the administration; the views expressed are solely those of the author.*

References

Carter, J. (1977). Environmental Message to Congress, The White House, Washington, D.C., May 23, 1977.

Crenson, M. and B. Ginsberg (2002). *Downsizing Democracy: How America Sidelined Its Citizens and Privatized Its Public.* Baltimore, MD: Johns Hopkins University Press.

IFMRC (1994). *Sharing the Challenge: Floodplain Management into the 21st Century.* A Report to the Administration Floodplain Management Task Force ("Galloway Report"), Interagency Floodplain Management Review Committee, Washington, DC.

Lindblom, C. (1959). "The science of muddling through." *Public Administration Review*, 19, 79-99.

Loucks, D.P., E. Stakhiv, and L. Martin (2000). "Sustainable water resources management." Editorial, *Journal of Water Resources Planning and Management,* 126(2), 43-47.

Maass, A. (1951). *Muddy Waters: The Army Corps of Engineers and the Nation's Rivers.* Cambridge, MA: Harvard University Press.

National Water Commission (1973). *Water Policies for the Future.* Final report to the President and the Congress of the US, Washington, DC, June 1973.

NCPS (2003). *Urgent Business for America: Revitalizing the Federal Government for the 21st Century.* Report of the National Commission on the Public Service ("Volcker Report"), Brookings Institution, Washington, DC.

NDPC (2000). *Preparing for Drought in the 21st Century.* Report of the National Drought Policy Commission, US Department of Agriculture, Washington, DC.

NPR (1993). *From Red Tape to Results: Creating a Government that Works Better & Costs Less.* National Performance Review, Vice President Al Gore, The White House, Washington, DC.

NRLC (1992). *America's Waters: A New Era of Sustainability: Report of the Long's Peak Working Group on National Water Policy.* Natural Resources Law Center, University of Colorado, Boulder, CO.

PCSD (1996). *Sustainable America: A New Consensus.* President's Council on Sustainable Development, Government Printing Office, Washington, DC.

Rogers, P. (1993). *America's Water: Federal Roles and Responsibilities.* Cambridge, MA: MIT Press.

World Bank (1993). *Water Resources Management.* Policy paper, The World Bank, Washington, DC.

WWPRAC (1998). *Water in the West: Challenge for the Next Century.* Western Water Policy Review Advisory Commission, NTIS, Washington, DC.

Vulnerability to Terrorism: Addressing the Human Variables

William D. Rowe, PhD[1]

Abstract

The terrorist attacks of September 11, 2001 revealed the vast scope of our vulnerability to terrorism. The reaction, and even overreaction, to these threats has led to costly disruption of both infrastructure and commerce, as the chaos in the airline industry so vividly displayed. A systematic framework is needed for analyzing the terrorist threat realistically and for coordinating the interaction among the responding organizations as well as the public.

What has yet to be assimilated about this risk from terrorism is the human variable. Whether the source of terrorism is international like al Qaeda or domestic like Timothy McVeigh, terrorists learn from experience and alter their tactics constantly. Accordingly, the measures necessary to protect people and property must be constantly reevaluated and revised, as in a never-ending "game." To be predictive in this environment, new tools are needed to properly assess the vulnerability to terrorism of systems and assets. New methods are required for determining who and what is at risk, how great the risk is, what measures can be taken to minimize that risk, and at what costs. In this paper, a simple game structure is described.

Introduction

Traditional methods of historical analyses, probability calculations, and actuarial tables are inappropriate to assess the risk of terrorism. These methods worked for floods, hurricanes, tornadoes, and earthquakes, since these disasters are governed by the unchanging laws of nature. However, terrorist motivations are varied and complex. Terrorists shift political goals and strategies as they learn from world events and their own experience with countermeasures. Risk management must move beyond the static world of historical probabilities to dynamic tools based on the principles of conflict among thinking opponents. One example is what mathematicians call "game theory."

---

[1]Rowe Research & Engineering Associates, Inc., 130 Tull Place, Alexandria, VA 22304; 703-929-8542; wrowe10@comcast.net.

While game theory has provided some structures, it has had limited application due to the complexity of even simple games. With terrorism, the rules of the games are constantly changing because thinking opponents look for new and existing vulnerabilities in the infrastructure, taking into account the limitations of resources for defense.

Simple Game Structure

This paper describes a simple game structure that demonstrates the motivational aspects that must be addressed in a more complex manner as well as the needed capabilities to carry out attacks. The motivations involved are as follows:

- Terrorists' positive motivations—the results of a successful attack.
- Terrorists' negative motivations—the things they fear.
- Defenders' positive motivations—prevent or avoid an attack.
- Defenders' negative motivation—minimize expenditure of resources.

Additionally, the terrorists must have the resources, capabilities, and opportunities to carry out an attack, and the defenders must "harden" their facilities to withstand such attacks, which includes lowering their profiles in the eyes of potential attackers.

In order to provide insight into the complex game and as a guide to understanding basic principles, a simplified game is presented in Figure 1. This figure is a *truth table* showing attackers' (terrorists) options in the leftmost column and the defenders' (targets) options in the top row across.

In order to damage a target, the attacker must be highly motivated and have the capability to successfully carry out the plan. In this simple example, the attackers are limited to four alternative strategies:

1. *Attack:* High Motivation, High Capability—attackers are highly motivated and have a high capability to carry out the attack successfully.
2. *Attack:* High Motivation, Moderate Capability—attackers are highly motivated and have a moderate capability to carry out the attack successfully.
3. *Don't Attack:* High Motivation, Low Capability—attackers are highly motivated but have little capability to carry out the attack successfully.
4. *Don't Attack:* Low Motivation, High Capability—attackers are not motivated to attack although they have the capability to carry it out successfully.

The defender has three alternative strategies:

1. Defend with high capability.
2. Defend with low capability.
3. Don't defend.

Game Outcomes

The alternatives of both the attackers and the defenders intersect to define 12 outcomes as shown, numbered I to XII. Should an attack be made, it is assumed the attacker wins if its capability exceeds that of the defender (dark shading). The defender wins in the reverse case (no shading); if they have the same capability, it is a toss-up (light shading). If there is no attack, then the defender wins.

The utility functions for each of the twelve cases are shown in Figure 2 for both the defenders and attackers, with the shading from Figure 1 carried over. From left to right, the first three columns show the situations, the box numbers from Figure 1, and the winner, either the attacker or the defender. The next two columns show the defenders' costs for protection—i.e., for extending the capability to defend—(High, Moderate, Low, Too High) and the costs of asset destruction due to a successful attack (High, Low, None), respectively. It is assumed that destruction costs would be very much higher than protection costs should an attack be successful. Losses to attackers are shown in the next two columns. The attackers lose face (No, Yes) both as individuals and their support groups, and may die or be captured (Some, High, Low, No).

Figure 2 also shows the alternative outcomes for two toss-up cases. Either the attacker wins (A-Win) or the defender wins (D-Win). The utilities change with the different outcomes. If the defender wins, these cases represent cost-effective deployment of defense resources. Otherwise, we can conclude that the defense costs were a bit short of what was required to ward off the attack. This is summed up as follows:

Cases	Defense Expenditures
I, V	Toss-up
II, III, VI	Inadequate
VI	Cost-effective
X	Over-expenditure
VII and XI	Cost-effective
VII, IX, XII	Either cost-effective or lucky

Discussion

This simple game illustrates how the motivations of both attackers and defenders and the capabilities to attack or defend shape the outcome of an attack. The attackers are motivated to cause destruction. The defenders are motivated to use scarce resources effectively. Different motivation and resource requirements show that this is an asymmetric open-ended game for which classical game theory has limited coverage.

The likelihood of an attack is not a probabilistic function, it is behavior-driven. The purpose of the game is to illustrate this point, and provide some insight into how this type of behavior can be used to address the terrorism problem.

Truth Table	Defenders Alternatives		
Truth Table for the Game	a. Defend High Capability	b. Defend Low Capability	c. Don't Defend
Attackers Alternative Strategies			
1. Attack Target - High Capability & High Motivation	Toss Up A	Att-Suc	Att-Suc
2. Attack Target - Low Capability & High Motivation	Def-Suc 5	Toss Up B	Att-Suc
3. Don't Attack - High Motivation	Def-Suc 1	Def-Suc 2	Def-Suc 3
4. Don't Attack - Low Motivation	Def-Suc 4	Def-Suc 2	Def-Suc 3

Figure 1. Game theoretic structure—simplified version.

Utilities		Box	Outcome	Costs To Defenders		Costs to Attackers	
				Asset Protection Costs	Asset Destruction Costs	Lose Face	Die or Capture, etc.
Toss Up		I.	A-Win	High	High	No	Some
			D-Win	High	Low	Yes	High
Toss Up		V.	A-Win	Moderate	High	No	Low
			D-Win	Moderate	Low	Yes	High
Attacker Success		II.	D-Win	Moderate	High	No	Low
Attacker Success		III.	D-Win	Low	High	No	Low
Attacker Success		VI.	D-Win	Low	High	No	Low
Defender Success		IV.	D-Win	High	Low	Yes	High
Defender Success		VII.	D-Win	High	None	No	No
Defender Success		VIII.	D-Win	Moderate	None	No	No
Defender Success		IX.	D-Win	Low	None	No	No
Defender Success		X.	D-Win	Too High	None	No	No
Defender Success		XI.	D-Win	Moderate	None	No	No
Defender Success		XII.	D-Win	Low	None	No	No

Figure 2. Utility functions for the twelve boxes in Figure 1.

SESSION 3

LESSONS LEARNED FROM EXPERIENCE DEALING WITH RISKS OF EXTREME EVENTS: PART I

Rapporteur: Ruth Y. Dicdican[1]

The five speakers in this session talked about a wide variety of topics, including schedule dependencies in hurricane preparedness, knowledge mapping across networks, extreme events in dam safety risk assessment, uncertainty calculi, fuzzy logic, and internet incidents. James Lambert of the University of Virginia chaired the session and introduced each speaker.

James Lambert *(University of Virginia)* described a method he developed with Clare Patterson that is used by a transportation agency to prioritize schedule dependencies in hurricane recovery. The approach involves identifying dependency scenarios where one agency is waiting for inputs from other agencies, and subsequent delays or risks slow down post-disaster recovery. The dependency scenarios are categorized according to the functional lines of the various agency units involved in the recovery effort. To minimize negative effects of schedule dependencies, analysis of dependencies can be performed in terms of the agency units involved, degree of dependency, and severity of the length of delay and time horizon.. A suggestion was given to further develop the tool and apply it to more events and scenarios so that adaptation and learning could occur.

Jim Lambert extended this method to terrorism risks to a highway, transportation, or communication network. He presented a method of network superposition for characterizing terrorism threats. The threat of terrorism has a hierarchical network structure. Therefore, one can juxtapose the threats to, for instance, a water network with the terrorism network, and identify the units that are more vulnerable to threats and the top occurring interactions. A meta-network called the *knowledge map* of terrorist threats to infrastructures can be developed by looking at information flows or relationships across networks. This knowledge map is used

---

[1]Graduate Research Assistant, Center for Risk Management of Engineering Systems, University of Virginia, Charlottesville, VA 22903; (434)924-0960; ryd3d@virginia.edu

for preparedness. A question arose on the loss of some information when using such a map. It was acknowledged that some information loss does occur, but the knowledge map is still useful for gaining information on scenario structures when the networks are juxtaposed.

David Bowles *(Utah State University)* talked about risk assessment of extreme events in dam safety. In estimating risk, one looks at the annual exceedance values from the site, regional data, and sometimes data gained from extrapolation. There is a credibility limit when one extrapolates. Beyond this limit, a prescriptive approach is usually followed. Risk is represented by the expected value. A question was raised on the validity of using expected value in dam failure when one is concerned with the tail of the distribution. When expected value is used, one has to exercise extreme caution since the order of magnitude of damages is three times higher than that of the benefits. Risk consequences are relevant to terrorism. Investments made to prevent these adverse events may also bring inconvenience to regular activities. One can make access to the dam controls harder, but this could also result in a longer time to perform regular tasks.. Such tradeoff issues arise in dealing with dam safety. The discussion that followed centered on these issues.

Scott Ferson *(Applied Biomathematics)* described the need for a difference in uncertainty calculus in the wake of terrorism. Uncertainty propagates through fault and event trees, arithmetic functions, "brave" extrapolations, and overconfidence of experts, among others. Uncertainty is represented by probability; however, *possibility* has nothing to do with probability. Scott gave an example of people flocking to buy lottery tickets when the prize money is higher, even though the expected value (or chance) of winning is now lower. People are attracted by the possibility not the probability of winning. Therefore, there is a need for a calculus of possibility. Traditional Bayesian models are not applicable because of the need for precise prior probabilities and the assumption of independence. A question was raised on the importance of prior probabilities in the presence of several likelihood functions. Priors become less significant, especially when stability is reached. Because terrorists are not normally distributed, distribution-free methods are needed, where one identifies the range of values without any underlying distribution. . Examples of the calculi for possibility include Aristotelian (modal) logic, worst-case analysis, fuzzy set theory, probability bounds analysis, and theory of surprise. A question was asked on the setting of bounds. Scott answered that if one does not know the bounds, one should make the range larger.

Istvan Bogardi *(University of Nebraska)* presented the use of fuzzy logic in risk analysis. Because of inherent uncertainty in the type of exposure (natural hazards or human-caused threats) and the threats to the system, probabilistic formulation may be difficult. Thus, fuzzy logic formulation is a recommended alternative. To perform analysis, interval analysis can be done with more available information sharpening the interval. Fuzzy numbers imply impreciseness and are also considered as multiple interval numbers A difference between fuzzy and probabilistic logic lies in the fact that although the fuzzy logic membership function curve looks like the probability distribution curve, in fuzzy logic the area under the curve has no meaning and thus, no probability implications. Fuzzy logic can be combined with probabilistic analysis. A simple example that dealt with risk analysis in infrastructure was presented to demonstrate the principles of fuzzy logic. The discussion centered on the meaning of the membership function and the concept that probability theory can also be used to solve the same problem.

Thomas Longstaff *(Carnegie Mellon University)* talked about the Internet meltdown. Some incidents have occurred, but in the context of a 100-year event, no such catastrophe (e.g., Internet meltdown) has yet happened. The closest that the Internet had was the "Code Red" worm that contained self-replicating malicious codes that attacked vulnerabilities in systems running the Microsoft Internet Information Server. The worm and its variants affected more than 250,000 hosts. Internet incidents are collected sets of events that violate a site's security policy. The count of incidents includes the international community, but events are US-based. The occurrence of incidents follows a seasonal trend. Most happen when school starts or ends and during holidays. The risk of a cyber-attack depends on the attractiveness of the site. A question was asked about the level of expertise that a terrorist must possess for the CERT Coordination Center to be interested in his activities. A fair amount of expertise is needed to attack the system. One has to understand the infrastructure and be well-funded to be able to plan and prepare carefully. A person with a graduate degree in computer science or one who has an undergraduate degree in computer science plus work experience and training has sufficient expertise. A question was raised on the capability of the government to protect itself from state-sponsored attacks. Intrusions usually need backdoor connections to the network. Thus, to attack a system, one has to understand networks well. With all the controls in place, attacks still occur because the system is largely controlled by lower-level people. The central part of the organization controls only a small part of the system.

SESSION 4

LESSONS LEARNED FROM EXPERIENCE DEALING WITH RISKS OF EXTREME EVENTS: PART II

Rapporteur: Joost Santos[1]

A. A Dynamic Risk Model for Information Technology Security in a Critical Infrastructure Environment
John Saunders, National Defense University

Key considerations in information security (IS) are:
Identifying threats and challenges;
Providing countermeasures;
Recognizing the role of temporal factors in the development of risk analysis frameworks; and
Modeling the dynamics of management decisions.

The speaker recommended a "must read" book relevant to IS:
McLure, S., J. Scambray, and G. Kurtz, 2001. *Hacking Exposed: Network Security Secrets & Solutions, Third Edition*. Berkeley, CA: Osborne/McGraw-Hill.

For every known IS vulnerability, there is a matching countermeasure. However, we do not know all IS vulnerabilities. The following tasks are geared towards identifying IS vulnerabilities, which ultimately can aid in pinpointing the potential threats and challenges to IS:
Improving the response time (alert services and user awareness) possibly via increased use of SCADA and standardized platforms,
Updating the procedures and educating the workforce, and
Identifying the "unknowns" via defense-in-depth strategies highlighting the importance of IS resources (people, processes/operations, technology).

---

[1] Graduate student, Department of Systems and Information Engineering, and the Center for Risk Management of Engineering Systems, University of Virginia; 434-924-3803; jrs8e@virginia.edu.

An effective IS risk framework must be able to address and convey the questions of risk assessment and management. Hence, it needs to stress the importance of identifying dynamic factors such as:
> Distribution of attacks over time;
> Frequency of patch-updates;
> Rate of software/hardware obsolescence; and
> Component failure rate (e.g., MTBF—mean time between failures).

To effectively evaluate IS decisionmaking policies, one must be cognizant of the trade-offs between risk and resource-allocation issues (e.g., how does risk improve with an increased level in IS investment?).

B. Quantifying and Communicating Model Uncertainty for Decisionmaking in the Florida Everglades
Pete Loucks, Cornell University

The speaker introduced the CERP (Comprehensive Everglades Restoration Plan)—a framework and guide to restore, protect, and preserve the water resources of central and southern Florida. The increasing human habitation in Southern Florida delivers serious perils to the Everglades. Construction of canals, levees, etc. interrupts the Everglades' natural sheetflow, which consequently diverts the needed freshwater to the sea. The plan requires a $7.8 billion investment and some 20 years to complete.

Restoring the Everglades requires extensive use and development of models to aid in the decisionmaking process. As with any modeling tasks, output errors can be attributed to uncertainties in inputs and other model parameters. Sources of such uncertainties include lack of knowledge and natural variability (e.g., hydro and meteorological data). Other sources of uncertainties stem from the model choice itself and errors in the model's underlying algorithms. Through simulation, an analyst can compare the model's predictions with actual observations. This allows model calibration, or shifting to a more appropriate model.

Distributions of performance measures (in contrast with point estimates) are appropriate to factor uncertainty into the modeling process. Such distributions enable the identification of both the range of values for a given performance measure and correspondingly, the likelihood of exceeding the perceived tolerance levels. For example, phosphorus concentrations exceeding 10 ppb could cause the conversion of sawgrass prairies to cattail marshes, which ultimately can eliminate the habitat for birds and fish. There are multiple performance measures in the Everglades restoration plan. Thus, conducting trade-off and sensitivity analyses are necessary for determining the performance targets. The analyst has to show the burden of proof for such performance targets (e.g., is there enough evidence to show that the critical level of phosphorus concentration is indeed <10 ppb?).

It is important to address the multiplicity of stakeholders. How can policymakers effectively communicate restoration plans to the affected stakeholders? A case in point—commercial fertilizers that are necessary in farming are known to contribute to higher levels of phosphorus. Thus, lowering the phosphorus concentration to 10 ppb or less would work at the expense of retiring the farming activities surrounding the Everglades. There might be a need to review the national standards from a cost-benefit-risk perspective. It might be more feasible to set a performance goal based on a known reference point (e.g., the status quo). For example, improvements in a performance goal can be measured from the *status quo* (e.g., 40% phosphorus reduction on Chesapeake Bay), rather than setting a strict *a priori* target (e.g. <10 ppb). Such a strategy can be implemented on incremental levels, thus allowing for policy adjustments grounded on "acceptable levels" of cost, benefit, and risk.

C. Game-Theoretic Models for Critical Infrastructure Protection
Vicki Bier, University of Wisconsin-Madison

Game theory models can be applied to Critical Infrastructure Protection (CIP). They:
 protect against intelligent and adaptable adversaries,
 recognize defensive strategies by taking into account attacker behavior,
 can identify qualitative properties of optimal solutions, and
 although beginning to be used in military IS, most CIP-related applications are still exploratory.

The speaker showed two basic game theory topologies for CIP: series and parallel models. The objective of such models is to identify resource allocation policies to minimize the "costs" associated with defending critical infrastructures. The functions in the currently proposed game-theoretic models are assumed to be convex, increasing, twice differentiable, and invertible.

For a system with parallel components, the derivative of the objective function can point out cost effectiveness. Thus, components that are "too costly" to defend would not be strengthened. On the other hand, all components in a serial system offer equal payoffs to the enemy, thus the defender must equalize the "success probabilities" for all the target components.

Cost encompasses not only the installation of resources to harden the system's components, but also the mitigation costs in the aftermath of an adversarial attack. The issue of single objective optimization (via assignment of weights to multiple objectives) was raised: not all real-world objectives are commensurable. The speaker justified that the proposed game-theoretic models utilized the multi-attribute utility theory (e.g., Keeney, R. and H. Raiffa, 1993. *Decisions with Multiple Objectives: Preferences and Value Trade-offs*. New York, NY: Cambridge University Press).

The two boundary assumptions on information availability (perfect knowledge and no knowledge) have drastically different implications for the modeling process. The *perfect knowledge* assumption may not be too unrealistic because we are an open society, not to mention the vast availability of public-domain information on the Internet. At the other end of the spectrum, the *no knowledge* assumption implies that the adversary selects a system component to attack at random. Higher adversarial knowledge on the vulnerabilities of a critical infrastructure intuitively poses more of a challenge to the flexibility of the defender.

Future efforts on game-theoretic modeling for CIP include relaxing the assumptions used in the current models, incorporating temporal issues (i.e., dynamics), and considering topologies other than systems with purely parallel and series components.

D. On Applying the General Theory of Quantitative Risk Assessment (GTQRA) to Combating Terrorism
Stan Kaplan, Food Safety and Inspection Service, USDA

Quantitative risk assessment answers the triplet questions: "What can go wrong?;" "What is the likelihood?;" and "What are the consequences?" The speaker mentioned the *principle of inversion* (see Anticipatory Failure Determination^{TM} in www.ideationtriz.com) in which the aim is to produce the failure intentionally instead of hypothesizing how the failure has occurred or shall occur. This is an innovative way of "converting harm into benefit." Instead of asking, "What can go wrong here?," the inverted question, "What can I make go wrong here?" can be more effective in identifying the initiating events which can trigger system failure. A failure scenario (S_i) is a departure from a system's defined *as-planned* scenario (S_0). Terrorism-based event trees and fault trees can be constructed for each identified failure scenario, thus pinpointing the possible surveillance points in the system.

Quantitative Risk Assessment (QRA) can be employed to assess the vulnerability of our food supply sector. The possibility of terrorist-contaminated meat production is an example of a safety concern in this sector. From a terrorist's perspective, planning a successful contamination of a food supply entails the following considerations:

Type of "poison" to use (e.g., chemical, biological, etc.);
Possible ways to obtain the poison (buy, self-manufacture, etc.);
Possible ways to access the food production plant;
Point of insertion (i.e., in which part of the food-processing chain can the poison be best introduced?);
Likelihood of the poison being noticed/discovered (immediate vs. delayed);
Introduction of poisoned food into the marketplace;
Human consumption; and others.

The following are the key discussion issues that resulted from the presentation:
Surveillance vs. defense of a vulnerable asset;
Feasibility of protecting myriad assets;
Varying motivations behind terrorism;
Psychological impacts of terrorism (in addition to physical); and
Distinctions between natural and accidental events and terrorist attacks.

E. Perspectives on the Characteristics of the Vulnerability of Water Resource Systems to Nature and Terror
Nick Matalas, Consulting Hydrologist

Potential terrorism scenarios of water resource systems:
    Hazardous materials (Hazmat);
    Chemical and biological substances;
    Explosive devices; and
    Destruction of bridges or locks.

Elements of vulnerability:
    Accessibility of water resources;
    Hydrological disconnect;
    Water navigation management on regional scale; and
    Aging of other related infrastructures.

The vulnerability of water resource systems can be reduced via hardening:
    Redundancy, robustness, and resiliency (3R's);
    Managerial connectedness (local, regional, and national); and
    Minimizing the likelihood of surprises.

Schacklean surprises are psychological in nature and categorized either as counter-expected or unexpected events. *Fiering-Kindler* surprises are categorized as structural, imbedded, hydrologic, institutional, informational, demand, etc. The various types of possible physical and cyber-based terrorist attacks on water resource systems were discussed. There was a suggestion to designate a coordinating agency to act as "data manager" for consolidating the risk scenarios in water resources. Sources of information for such risk scenarios can include the Defense Threat Reduction Agency (DTRA) and Office of Homeland Security, among others. The industrial and private sectors are also good sources for documenting risk scenarios; however, they are not likely to volunteer information, especially that of a proprietary nature.

SESSION 8

PANEL DISCUSSION: SYNTHESIS—WHAT DOES IT ALL MEAN?

Rapporteur: Jim Lambert[1]

Panel members were **Vicki Bier,** *University of Wisconsin-Madison;* **Scott Ferson,** *Applied Biomathematics;* **Yacov Haimes,** *University of Virginia;* **Stan Kaplan, FSIS—USDA;* **Nick Matalas,** *Hydrologist; and* **Gene Stakhiv,** *USACE.*

This final panel and group discussion revisited topics from all of the week's sessions. It was acknowledged that the conference has addressed the threat of terrorism to diverse systems and structures, including dams, ecological systems, electrical and computer networks, waterworks, and others. We considered that risk analysis must change in the face of terrorism; the fortress model is not sufficient, the trusted insider may no longer exist, the distinction of *us* versus *them* is evolving in its meaning and usefulness. Importantly for the quantitative analyses of terrorism, the modeling uncertainties associated with such paradigm shifts are overwhelming considerably the narrowly conceived distribution and parameter uncertainties.

Homeland security is fundamentally an intractable problem; the mind of a terrorist resists modeling. This week we have seen conventional methods applied to parts of the problem, and nonconventional methods applied to other parts. Perhaps we have merely succeeded in itemizing our paranoias, considering them in all combinations and from multiple perspectives. The CIA and the War College may be way ahead of us as risk analysts of these topics, on which they have been spending considerable intellectual and capital resources for over twenty-five years. It is astounding how vulnerable we are to cyberterrorism—a small computer and keyboard, at little expense, is able to cause damage consistent with the bombing of a power station. We must make appropriate investments to protect our soft sides.

---

[1]Research Assistant Professor and Associate Director, Center for Risk Management of Engineering Systems (CRMES) and Department of Systems and Information Engineering (SIE), University of Virginia, Charlottesville, VA 22903; 434-982-2072; lambert@virginia.edu.

This week, we have tended to frame issues in the language of conditional probability, proceeding to estimate a consequence given the occurrence of an event. That event (terrorism) has resisted modeling by probabilities. We have been led to examine the motivations of our antagonists, to ascertain emotions, to work with unprecedented uncertainties. But we must get on with it! We don't have twenty years to argue about terms and meanings.

If terrorism is difficult to fight, one alternative to fighting is to passively avoid the blows, while another is to preempt attacks through intelligence. The challenge is to use our imaginations in unfamiliar ways, to consider how vulnerabilities of engineering systems translate to the public at large. There is no guarantee that the uncertainties are sufficiently small to be relevant to action—decisionmaking may ultimately proceed from the heart and gut in particular cases. We can recall Sir Edward Grey stating at the outset of the first World War, "The lamps are going out all over Europe; we shall not see them lit again in our lifetimes." But is such a perspective too extreme for the current threat of terrorism? It is likely that we don't know what is the true threat, whether it's to cyber or water resources or some other.

The problem involves multiple dimensions, the threat is dynamic, and involves many constituencies. What have the systems theories of Wiener, Bertalanffy, et al. contributed in past decades? We must seek to identify the "systems" problems and recognize that good risk analysis is associated with good systems modeling. It will take considerable effort to return to the basics. We can look at terrorism and terrorists as systems, identifying short-, medium-, and long-term impacts and remedies.

We can consider whether to replace risk-cost analyses by vulnerability-cost tradeoff analyses, and we should not get hung up on estimating probabilities. We can do better if we coordinate quantitative with qualitative risk assessments, where quantitative assessments may be less effective for non-recurring threats. In this direction, we have seen risk represented as the cross-product of threat x vulnerability x consequence. We can elect, for example, to invest nothing on the threat and invest intensively on the others. We can develop scenarios without probabilities, adopting a gaming and contingency approach.

Consider that a talk this week assessed that there may be few long-term economic consequences from major terrorism events, or even long-term damage to infrastructure. Judging only by the economists' numbers, some could conclude that the loss of people is not a long-term or an economic loss, since human "capital" can be replaced in most cases. We can be sobered or outraged or enlightened from considering such analyses. We must be careful to consider what the local, state, and federal interests are in particular scenarios of terrorism.

We must consider that when we think about terrorism, we think very rarely about water resources. Why? There is considerable stored energy in dams and much potential to upset transportation by damaging river and lake locks. Why is water often not near the top of our list?

We have considered that in a systems approach, the relevant "state variable" of our nation may be the national soul, or will. Over several decades, risk analysis has been an exercise that eases our souls, as we take comfort in quantification. We have used frequentist, Bayesian, probability-bounds, fuzzy, and other modeling approaches to turn the unfamiliar (non-recurring) into the familiar (that which can be imagined as recurring). As the cold war "terrorized" us into doing good science in the 1960s, perhaps the threat of terrorists is catalyzing us to do good risk analysis at the start of the new millennium. We will do well to consider how risk-taking (in the face of terrorism and other threats) defines the national soul, which risks are honorable and energizing to the soul, and which are not. Consider going to the conference of the American Political Science Association; you will not emerge saying you can't model politics—quite the opposite!

Terror is one of a multitude of problems that are faced by water resources in the US and abroad. Let us not overemphasize terrorism in our planning—rather, let us get the emphasis just right. Let us make an educated judgment of which is more serious—the terrorist threat or aging pipes and valves. We need quantification to inform such judgment, but we may not invariably need probabilities—probabilities are less important than the undertaking of risk analysis.

SUMMARY OF RESPONSES TO PARTICIPANT QUESTIONNAIRES
1980 – 2002

On the last evening of each conference, a questionnaire was distributed to the participants. Recognizing that these are the tenth such proceedings, we have compiled the responses to the same three questions from all ten conferences. We have begun analyzing the trends that emerged, and the results should make an important contribution to the field of risk-based decisionmaking in water resources.

The following questions were asked at each conference:

1) List the three most important issues/aspects/elements related to risk-based decisionmaking that were raised during this conference.
2) List the three most important new ideas/concepts that you learned during this conference which would be helpful in your job.
3) List the three most important issues needing further study in risk-based decisionmaking in water resources.

An edited comprehensive list of the responses from the last ten conferences follows:

1. The most important issues/aspects/elements related to risk-based decisionmaking:

1980

- Framing the proper questions for application of risk/benefit analysis.
- Making explicit any exercise of value judgment.
- Translating the results into laymans' terms for use by the decisionmakers.
- Quality control of risk/benefit analysis.
- Groundwater contamination as a candidate for the successful use of risk/benefit analysis.
- Probability of failure of major structures.
- Confidence bonds for project benefits, project costs—capital and operational, and probability of specific environmental impacts.
- Understanding the data requirements for risk assessment.
- Analysts'/decisionmakers' understanding of public perceptions, needs, and preferences concerning the reliability/risk associated with projects.
- Defining clearly, for each situation, the objective one seeks; i.e., is it the need to evaluate extreme events (via risk/benefit analysis) or to establish probable outcome (risk and uncertainty)?
- Need to implement the idea of risk and uncertainty in project formulation in order to establish the habit of thinking in terms of probable outcomes.
- Risk estimation techniques.
- Risk/benefit tradeoff decision process.
- Need for a clearinghouse that collects information on events subject to risk/benefit analysis and makes data or analysis available to practitioners.
- Need to very clearly differentiate between "risk" and the use of certain data—which should be referred to in terms of its degree of reliability.

- Need to convey reliability of the data used to planners, the public, the decisionmakers, etc.
- The social and institutional barriers to implementation of results.
- Importance of simplicity for the effective utilization of risk/benefit techniques.
- The persistence of value judgment as the dominating factor in water resources decisionmaking.
- The equal importance of the process and the methodology in risk/benefit analysis.
- Need for risks to be explicitly identified and addressed, rather than hidden in subjective judgments.
- Need for improved communication among water resource analysts as well as between analysts and lay decisionmakers.
- Lack of objectivity of ultimate decisionmakers, who are concerned mainly with getting the most for themselves or their constituents.
- Developing practical, simple methodologies for risk/benefit analysis.
- Developing practical interactive multiobjective methods which display the alternatives in a way that is easily perceived by the decisionmakers.
- "Optimizing" the operating and planning of water systems wherever reasonable estimates of probabilities can be made.
- The role that risk/benefit analysis could potentially serve in communicating the planning process.
- Relating risk assessment to safety factors in engineering design.
- Catastrophic risk.

1985

- The necessary integration of the value of human life into dam safety concerns.
- The apparent lack of agreement between "theoreticians" and "practitioners" regarding risk procedures and their utility.
- The need for consensus regarding the probability of remote events such as the probable maximum precipitation (PMP) or probable maximum flood (PMF).
- The difficulty of dealing with noncommensurate risks (among the alternatives).
- Simple models do not accurately reflect the way decisions are made.
- Acceptable risk level is dependent on the alternative solutions.
- Attempts to avoid legal liability cause many decisions to extend beyond what is "reasonable."
- Program decisionmakers do not understand the public's willingness to accept risk in the varied areas which threaten their lives individually or collectively.
- The priority on model development over improving databases for input makes real progress (in risk analysis) slow.
- Cost of standards; data to support [standards] and who pays [for conservative standards].
- "Nonstructural" approaches to reduce or pay for costs of failure.
- Institutional means to deal with uncertainty based on assessment of consequences.
- There is a huge resistance to moving away from standards.
- Decision consequences are more important than probabilities.
- Systems under consideration are poorly defined.
- Identification of barriers to more widespread use of risk-based analysis in design decisions (more costly design process, need for new expertise, liability, institutional rigidities, etc.).

- Ways to overcome such barriers (modifications to codes; agency standards and design procedures; training; new laws limiting liability; financial incentives to grant programs; statements of congressional policy (RCRA model); reducing perceptions of liability exposure; publishing new and innovative applications of risk analysis in design; etc.).
- [Individual/professional] legal liability if one deviates from standard practice.
- Risk analysis is beneficial to decisionmakers.
- Acceptable levels of risk depend not only on individual concerns and situations, but also on time (hindsight).
- Value of human life.
- Identifying and quantifying variables to be analyzed over future time in risk-based decisionmaking.
- Standards vs. risk-based decisions and legal and moral liability.
- Liability of engineer in using design less conservative than standard, and what motivation (or incentives) there are (i.e., sharing in benefits and/or having client accept liabilities) in applying risk analysis.
- Multiobjective decisionmaking—e.g., the value of human life, preference modeling in such cases, and assigning responsibility for decisions (engineer or owner).
- Case studies [are valuable and important].
- [Legal] liability for using risk analysis vs. standards for design (to the designer).
- Acceptability of using risk analysis [in terms of] cost savings vs. standard "conservative" designs.
- The cost of "human lives" and consequences in using risk analysis vs. standard designs.
- Issue of how to analyze risk when probabilities are poorly defined.
- Use of risk analysis to define (or redefine) engineering standards.
- Idea of maximizing or minimizing the proper objective.
- The need to develop a unified approach to risk analysis.
- Risk-based analysis is being used in design of water resource systems. Significant advances since the 1980 conference (*Risk-Benefit Analysis in Water Resources*).
- The concerns about liability as it discourages risk analysis (not necessarily a bad thing).
- Validity of probabilistic approaches (in risk analysis) which depend on assumed distributions and limited databases.
- Gap between researchers, practitioners, and decisionmakers in understanding and applying risk analysis.
- Limited application of risk analysis concepts to other water resource problems besides dam safety.
- Risk analysis is a necessary element in water resources planning, design, and/or operation.
- Diverse methods that are available for risk analysis—it is important to use the most applicable methods.
- A possible rational method to analyze alternative plans on some basis other than established standards. Standards are generally not arbitrary and have evolved because of social, political, or environmental reasons.
- Concept of liability.
- The problem with treating rare events analytically.
- Multi-attribute [approaches to] decisionmaking.
- Policy capturing—(conjoint analysis).
- Role and responsibility of engineers in social policy decisions.
- The unending difficulty in setting standards.

- Role of insurance—its potential for managing risk.
- Essential differences between structural failure and hydrologic failure—require different standards.
- Risk shifting and/or risk sharing as a strategy of dealing with residual risk.
- Economics of engineering design and control of risk.
- Risk-based decisionmaking is an approach to yield better decisions. It doesn't make decisions.
- Problems of incorporating loss of life into risk analysis. The major problem!
- Engineering standards need to be evaluated periodically—risk analysis can help.
- Linear utility assumption required for risk analysis (expected values).
- The usefulness of risk analysis in engineering design and decisionmaking versus the indiscriminate application of standards.
- The differentiation of risk analysis from other forms of systems approaches as a continuum of necessary evaluation approaches.
- The implicit or inherent statement of social risk acceptability contained within a standard.
- The critical role of "expected value" and uncertainty of probabilistic statements in risk analysis.
- Actual utilization of [risk] analysis by decisionmakers.
- Confusion caused by the value of "human life" in evaluation of alternatives. We must deal with it or close shop.
- Is liability a real issue for engineers? Where is the data?
- Not everyone has same expectations, concepts on methods for risk assessments.
- Purpose [of risk analysis] is to define basis of negotiation; it is one tool in the decisionmaker's tool box.
- Uncertainty exists in data to the extent that a risk analysis is not always preferable.
- Importance of reliable data (probabilistic information) in performing risk analysis.
- Need to identify decisions, analyze consequences and the risks.
- Need to properly communicate results of risk assessment to decisionmakers in a clear and understandable format.
- Increasing liability [with use of risk analysis].
- Need for conceptual/methodological consistency [in application of risk analysis].
- Practical dilemmas of decisionmaker.
- The need for more understandable and implementable risk-based assessment techniques.
- Where does the responsibility of the engineer fit in guiding the decision process?
- The inherent weakness of the engineering community to deal effectively with very-low-probability events.
- Variability of data at the extremes.
- Availability of mathematical methods but lack of extensive database.
- Political process as the ultimate decision network.
- Risk/liability.
- Risk analysis is tool, not end or religion.
- Variety of tools for risk analysis.
- What procedures are available for assessing subjective uncertainties?
- Implicit consideration of risk in design standards.
- Sequential decisions.
- Data-gathering.

1989

- Expert opinion doesn't necessarily predict actual problem distribution.
- Yarning systems may be more cost-effective in mitigating losses than engineering fixes.
- Risk communication is as important as (or more important than) the choice of analysis methods.
- Use of expert(s) in risk assessment.
- Role of perceived risk in behavior.
- Should risk of extreme events be addressed with a separate calculus?
- What is role of political process in establishing risk-based criteria?
- Risk analysis is an art, not a science.
- Validation of expert judgment.
- Modeling without using real data to calibrate is useless.
- Risk analysis is overly quantified without giving thought to nature of decisions needed.
- Decomposing risk problems.
- Model structure/formulation.
- Theoretical description beyond "common sense" considerations.
- Self-view of risk analysts: "benevolent expert" or public representative?
- The effect of risk perception on decision.
- The difficulty of making sound decisions based on quantitative risk analysis.
- The possibility of avoiding wrong decisions, based on risk analysis.
- How are risk criteria communicated to the public and to decisionmakers?
- Excellent papers that bridged the gap between engineering, economics, and social sciences.
- The "decisionmaker" was addressed with some detail regarding skills, experiences, place in the organization, and used to either explain (Goicoechea) or predict (Baumann) decisionmaking behavior.
- How to integrate public perceptions and expectations into risk-based decisionmaking.
- Utility of fuzzy sets and other uncertainties in modeling.
- Utility of ecological models.
- Difference in risk criteria/levels of acceptability between various specialties.
- Definition of alternative actions for hazard mitigation (waste cases).
- Probabilistic, fuzzy, and mixed approaches to risk assessment.
- Multicriterion decisionmaking.
- Importance of expert judgment.
- Possible use of fuzzy sets.
- Abuse of risk-based decisionmaking.
- Risk assessment of dam safety.
- Overestimation of impact of climate change.
- Use of methodologies for cases.
- Communication of risks/benefits.
- Use of risk analysis to mediate conflicts and minimize social costs.
- Importance of process in risk decisions.
- Empirical importance of informed consent.
- Role of models in ecological risk assessment.
- Techniques of risk assessments for human health and ecological resources.

- Bureaucratic tendencies for detail without considering broad issues and assessing a broader set of options.
- Conceptual errors in use of expert opinion.
- Shortcomings of using simple expectation in risk analysis of extreme events.
- Fuzzy risk analysis.
- Risk communications.
- The learning function of case studies and personal experience in risk management.
- Integration of "objective" risk estimates with the overall decisionmaking process that may be subjective.
- Interface between the public and the technocrat.
- Bayesian approaches and extreme-event analysis.
- The effect of risk perception on decisions.

1991

- Use of fuzzy-sets theory for risk analysis including selecting probability levels.
- Data requirement and its relation to approaches adopted, Bayesian approach, or fuzzy-set theory.
- Large-scale infrastructure problems in risk-based decisionmaking.
- Decisionmaking using risk information is much less developed than risk assessment methods.
- Engineers still reinventing economic theory.
- Use of risk analysis for practical management of water resources.
- Different ways of expressing risk and reliability measures for use in water resources.
- Risk communication (education, management, conflicts).
- Chaos between public policy and risk-based decisions in many areas.
- Gap between infrastructure needs and current national policy.
- Movement of engineers toward probabilistic models.
- Is "uncertainty" equivalent to probability?
- Does risk analysis equal probability?
- Use of forecasting.
- Methodologies are applicable across a wide range of problems in water resources and other areas of public works.
- Infrastructure rehabilitation is a promising area of research for risk analysis.
- There is a serious problem of educating students at undergraduate level to understand statistics and risk analysis.
- Clear distinction between uncertainty and risk.
- The increased importance of the political process in managing water resources systems.
- Refining the formalism of risk analysis using Bayes' theorem.
- How to elevate the most critical cost vs. risk tradeoffs for decisionmakers.
- How to isolate the critical cost/risk decision variables from all other uncertainties.
- The seeming lack of use of the work done to-date. (Are we merely satisfying ourselves rather than others?).
- The conflict between probabilistic versus fuzzy set approaches.
- The need for careful problem/system definition and the impact of intended solution methodology on that process.

- Integration of risk (or measurements) with multiple objective analyses—how, when?
- Practical applications of fuzzy-set methods.
- Implementation of risk/cost philosophy into national policy decisions.
- Broad education of risk analysis thinking into society activities—public/political value selections.
- The role of Bayes' theorem in risk assessment, decisionmaking, and decision implementation.
- The wetland dilemma.
- Conflicts in health and safety matters.
- Fuzzy sets do not seem to reduce fuzzy thinking.
- How valid is the judgment of experts? How do we test it? How do we validate good judgment?
- Many of the mathematical, structural approaches presented seem to miss the real-world behavior found in the problem environment.
- Relationship between risk analysis and engineering standards.
- Risk analysis of environmental issues (water pollution, hazardous wastes).
- Rehabilitation of existing infrastructure (risk analysis and assessment).
- Stan Kaplan's observations about the body of theory he called "quantitative risk management" confirm my pre-existing belief that we have a framework/specialty here that deserves to be formalized and taught in an interdisciplinary manner at more universities on a regular basis (i.e., a course suited for undergraduate engineers, economists, business, decision theory, and other majors). N. Dudley's "multiple single-purpose reservoirs" model is an experiment that merits watching, though the risk dimension of this was somewhat incidental.
- There is a great need to define engineering standards and to identify those that can be replaced with risk analysis. Project planning and financing of infrastructure, major rehabilitations, reservoir operation, and a variety of themes dependent on our capabilities.
- The combined use of fuzzy-set and probability distribution is feasible for some cases.
- Communication between analysts and decisionmakers plays an important role in the decisionmaking process.
- The effect of human behavior on decisionmaking.
- Global climate change.
- Engineering discipline's requirements.
- Refurbishment of infrastructures (setting priorities).
- Policies of drought management.
- Need for good organizational and institutional incentives to promote effective risk-based decisionmaking.
- The basis and reasons why different approaches are taken for highly uncertain health risks versus better understood safety risks.
- Development of new tools for estimating the value of information.
- The new concept of probability and utility raised by Professor Pate-Cornell.
- Wetlands—policies and issues.
- Many engineers and decisionmakers lack formal statistics and probability courses (clients need to understand concept of risk analysis and management).
- Need to change culture from standards-based process to risk-based process.
- Risk-based decisionmaking is not unique to water resources.
- What is the value of expert opinion?
- The importance of holistic formulation of problems.

- The need to educate engineers with the objective of giving them the ability to synthesize, work in teams, and think in terms of systems.
- Risk related to climate change.
- Infrastructure refurbishment.
- Wetland-related risk.
- People define risk and risk-based decisionmaking in different ways.
- The difficulty is primarily in measuring.
- We need to reach out and bring in more industry, managers, and analysts.
- The new Stanford work on uncertainties of probability distributions.
- The work of Jennie Rice for hydro and environmental planning.
- Historically developed engineering standards produce very different and uneasy levels of "safety" or risk evidence when evaluated in terms of qualified risk.
- Uneven approaches to risk assessment and risk perception lead to major blunders of public and private policy in risk evidence and related use of resources.
- Another dimension of risk management objectives was suggested in that individuals appear to have different levels of aversion to risks arising from randomness versus lack of information.

1993

- Apostolakis's formulation of model uncertainty using subjective likelihood functions: P_r[satisfactory assumptions/evidence and objective].
- The semantic disagreements and misunderstandings of the meaning of risk and probability are still unresolved.
- Risk-based formulation of 0 and M problems (dredging, repair policy); what should be the decision criterion?
- Need to start thinking about a new "social risk compact" between the public and engineers to clarify the limits of social risk acceptance.
- Risk analysis now becoming a part of agency analytical procedures.
- Still a controversy between statisticians: Bayesians and fuzzy-set theorists.
- De-biasing incident database, introduced by Pate-Cornell.
- Concept of "poorer is riskier" advanced by Dick Schwing.
- Talk by Tony Thompson (a lawyer) on the federal risk management policy or the lack thereof.
- Communication to participants in decisionmaking processes.
- Structure of decisionmaking process needs to be developed to use risk infrastructure as a consensus-building process.
- Still need to gain acceptance of risk analysis from professionals and managers who think deterministically.
- How to get probability distributions when you don't know much.
- Comparison between present-worth models and other models.
- Lack of criteria for selecting acceptable risk for decisions in public policy.
- Incorporate risk-cost analysis in engineering design.
- Difficulty in realistic estimation of probability distributions, e.g., cancer risk rates.
- Directions that the federal government is heading; highly diverse styles of EPA versus Corps in risk area.
- Methods of communicating uncertainty on decisionmaking.

- Impact of model uncertainty on decisionmaking.
- Need for the ability to demonstrate that risk and reliability can be applied to other fields.
- Methods for handling time preference.
- Questioning "standards."
- Model uncertainty and impact on decisionmaking.
- Risk models and case studies.
- Implementation of these ideas to a more general audience.
- Significant progress in risk balancing across specialty areas.
- Need to continue to link water resource research to other areas of risk analysis.
- Relating regulatory policy to risk-based concepts.
- Advances in including uncertainty.
- Organizational/institutional/public constraints in using risk-based concepts.
- Technology transfer: A great deal of effort is needed to transfer the knowledge about risk-based procedures to the engineering community.
- Adoption of risk-based design: methodologies have been developed that need to be used in practice.
- Engineering design standards: future engineering design standards need to be founded on risk-based standards.
- Use of risk analysis in engineering design will be hampered by lack of data in uncertainties and appropriate distribution to describe the uncertainties.
- Public's different evaluation of "everyday" risks to which they choose to expose themselves and risks that they feel are imposed on them.
- Environmental impact risk analysis.
- Commercial software for risk analysis.
- Rapid expansion in the federal government of risk-based decisionmaking that is relevant to a variety of water resources applications.
- Importance of training students in these areas of expertise.
- Monte Carlo methods for risk analysis are software-driven and often have nothing to do with reality.
- Focusing on uncertainty rather than risk is appropriate and healthy.
- Understanding uncertainty and how to harness it is an important new direction in the field.
- "Ecology risk assessment is back to square one."
- Implication for social decisionmaking of increased use of risk analysis (changing role of public and expert).
- Need to phrase risk in terms of "lives versus lives" instead of "lives versus dollars."
- The changing understanding of risk by our legal institutions.
- Public perceptions versus expert information.
- The tools are there; how do they get factored into the decision process?
- Costs of complying with "irrational" regulations.
- We still have a long way to go regarding acceptance by state and local agencies and citizens.
- General movement toward requiring risk-based decisions in federal agencies.
- Confusion still exists as to the difference between parameters and model uncertainty.
- Concern that the legal processes are so strong in the US that collection of anonymous data may be infeasible.

- Versatility of the risk analysis methods as developed for water resources applications, such as the application to anesthesia.
- Consumer/customer acceptance and understanding of models.
- Capability of analysts has advanced faster than capability of decisionmakers to process risk information.
- Risk communication: how can it be done effectively?
- Breadth of applications range for risk-based methods.
- Regulation-based risk control appears to be much less cost-effective than risk aimed at individuals or specific issues that directly affect people.
- Limitations of GCMs (Global Circulation Models) and the limited (or lack of) progress that has been made so far in the area of global climate change.
- Use of multiobjective decision trees in studying global climate change.
- Global climate change needs to be translated to regional impacts.
- Risk/decision analysis may not be helpful in climate change impact studies at this time; needs a lot more thought and research.
- Extent to which current water management practices may be adaptive to climate change challenges.
- Uncertainty in climate-change models casts serious doubt on their utility for forecasting future climate change.
- Banking approaches to balancing wetland and irrigation needs/objectives.
- Distinguish incremental and irreversible decisions in water resources.
- Relevance of possible climate change to water management (Stakhiv says "nyet").

1995

- The new legislative moves supporting risk and cost-benefit analysis.
- The poor state-of-the-art of GCMs and related modeling.
- The importance of trust as an aspect of risk assessment and management. The use of multiple failure modes in water distribution.
- The risk-reduction perspective of reducing lead and mercury in Michigan.
- Using extreme-event analysis to help aid in decisionmaking.
- The topic of risk aversion and how weakly- or strongly-averse attitudes set policy and regulations.
- Risk-based decisionmaking must be developed to incorporate some sort of risk benefit-cost approach in considering the end product. I
- There are practical ways to show risk-based/reliability assessment of engineering applications.
- Data quantity/quality problems.
- Precision vs. realism.
- Usefulness of fuzzy numbers.
- There seems to be a trend to use these techniques to replace a basic understanding of the actual science and engineering of the problem.
- Water resource engineering design and analysis measures need to be based upon standardized risk-based procedures.
- There seems to be a large disparity in definition for risk/reliability.
- The proliferation of risk-analytic approaches without standardization. No basis exists for choosing the "best," most valid methods for problem solving.

- The risk analysis profession is unprepared for dealing with the large-scale programmatic risk-benefit requirements of the new legislation.
- Risk communication methods do not seem to be having much effect on behavioral changes.
- Proposals and implications of pending federal legislation regarding risk assessment in connection with regulations.
- Appropriate types of problems for decisionmaking.
- Problems of communicating concepts to lay people who may need to participate in the process.
- Climate change has important implications for water resource management, especially with respect to variability.
- Importance of balancing considerations and extremes as well as central tendencies in design.
- Importance of using multi-function, multi-party views of water projects to achieve greater overall social benefits.
- The relationship between risk reduction vs. cost.
- The relationship between uncertainty reduction vs. cost.
- The realization that risk calculation and uncertainty reduction are big business. The field of study has taken on a life of its own for entertainment and profit. Studies take longer and cost more than fixing the problem.
- Fuzzy arithmetic as a tool for risk assessment and management.
- Ecological risk analysis.
- Risk-based pending legislation in Congress.
- Importance of new legislation on risk; Michigan doesn't seem to use it for mercury in water.
- Uncertainty analysis by probabilistic and fuzzy methods, especially for quality problems.
- Worthlessness of some government agency techniques.
- Convincing federal agencies to adopt risk-based decisionmaking.
- Collection of data needed for risk analysis.
- Lack of uniform understanding of definitions of risk.
- General lack of accessible data for risk analysis.
- Apparent lack of technology transfer to the user community.
- Variability in applying these concepts to activities within different federal agencies.
- Water-supply system reliability.
- Climate change.
- Legislative issues.
- Analysts are still accepting (in most cases) a threshold value and defining failure as exceeding the threshold. The threshold needs to be abandoned; i.e., a systems approach to examining tradeoffs needs to be adopted.
- Len Shabman's strongly-risk-averse/weakly-risk-averse dichotomy.
- Dan Willard's scale of stability concepts in ecological systems.
- Models for decisionmaking: there is still an issue as to whether our models should be predictive (accurate) or used to guide our thinking, with rules/decisions made based on "expert opinion."
- Do the decisionmakers want to know how uncertainty influences the results of our models? No, but they should. Therefore, there's a need to educate decisionmakers.
- The concept of standardization of risk assessments vs. that of site-specific or researcher-specific risk assessments.

- Legal/regulatory constraints to risk-based methods.
- Incorporating statistics of extremes.
- What is ecological restoration?
- Public perception.
- Lack of risk basis for EPA standards.
- The legislative initiatives which, if enacted, will require risk analysis to be performed, and the implications of this legislation.
- Risk evaluation—natural ecosystems.
- Extreme events: risk, the critical nature of extreme-event analysis, and methodology.
- Climatic changes.
- Relationship between risk, reliability, and uncertainty.
- Legislative issues on risk.
- Use of subjective probabilities.
- Extreme-event methodology.
- Need to use a complete risk analysis framework, don't ignore the benefit/cost component. Look at all sources of variability.
- Regulators' concern that risk analysis is an add-on to an already cumbersome process.
- Need to develop process-based continuous simulation models. Can't represent reality by regression equations and simple PDFs only.
- Decisionmakers' need to expand options in risk management decisions. Benefit-cost analysis needs to extend to options beyond our immediate department, agency, institution, or profession.
- Decisions usually have unintended consequences.
- Communication, problem-framing, and professional/personal biases are very important in risk-based decisions.

1997

- The use of expert opinion.
- Should we have definite standards in risk or not?
- Dam-safety issues.
- Safety levels for hydraulic structures.
- Some participants were uninformed about past efforts; some have retrogressed.
- New awareness in risk and uncertainty.
- Risk analysis is becoming overly institutionalized in federal agencies, especially EPA!
- How to bring risk assessment into the regulatory process.
- Growth of acceptance of risk analysis.
- Lack of government support for risk analysis.
- Comparative risk analysis was not discussed.
- Methods are developed between federal agencies without reading the literature.
- We have to learn more from each other.
- Necessity to manage uncertainty and consequences.
- Quantitative modeling in ecosystem protection and quality for risk analysis/risk management.
- Willful and purposeful hazards to water resources as a risk assessment/risk management issue.
- TRIZ. (2)

- Lack of multiobjective decisionmaking.
- Roles of uncertainty, variability, and bias in decisionmaking.
- Safety issues.
- PDF specification.
- Extreme-event analysis. (2)
- Time-dependent modeling needs attention.
- Models for evaluation are being developed.
- What is the value of judgment in risk-based decisionmaking?
- How do decisionmakers value risk-based decisionmaking?
- How does this information influence different kinds of decisions?
- Credibility does not come from science or rigor, so where does credibility enter?
- Communication of results needs work.
- Monte Carlo simulation.
- Time-dependent reliability and hazard function.
- Economic cost-benefit analysis.
- Risk aversion for extreme events.
- Fuzzy-rule-based modeling.
- Should risk assessment be primary task?
- The need for broader considerations.
- Notion of how safe is safe!!
- Risk and tradeoffs.
- Decision criteria.
- More explicit practical methods to express uncertainty vs. variability.
- Misunderstanding of ecologic risk assessment in real world.
- The attempts of multiple agencies to define acceptable risk as a benchmark.
- Multiobjective analysis/selling of risk analysis to the public.
- Comparing risks in daily life with extreme events; seeing how small they really are.
- Noticeable lack of improvement in risk analysis in the last decade.
- General applicability of fuzzy logic and sets to risk-based decisionmaking.
- Links to environmental risk assessment.
- New methods in handling uncertainty.
- Fuzzy sets as applied to water resource problem.
- We are falling behind!! Retrogressing simple mindedly...

2000

- *Risk communication*
    - The problem of language and terminology in risk analysis.
    - Need to link risk analytic technologies and findings to communications.
    - Risk communication and continuing harm of "100-year flood" terminology.
    - Decisionmakers and the public do not understand concepts of risk and probability.
    - Explaining risk to general public.
    - Being able to understand what information is needed by decisionmakers to help them make better decisions.
    - Risk analysis = risk assessment + risk management + risk communication.

- *Acceptable risk*
  - What is acceptable risk?
  - Persistent difficulty in defining acceptable risk as a meaningful concept.
  - Is acceptable risk a good concept?
  - Risk acceptance vs. risk tolerance.
  - The struggles of analysts and program managers to reconcile hard mandates with inherent subjective nature of acceptable level of risk.
- *Residual risk*
  - Residual risk is often a consequence of a chosen decision, rather than a specific desired outcome driving the decision.
  - Is calling it "tolerable risk" a sufficient improvement? Rather, should we view it as a residual risk after one has made decisions that were the best they could find?
- *Value/economic aspect of risk*
  - Connecting economics and risk analysis.
  - Engineering reliability analysis combined with economic risk-based modeling.
  - Engineering reliability and risk assessment, including economic analysis.
  - Environmental and economic risk issues associated with water resource projects.
  - Public agencies in US generally rely on prescriptive, single-objective, economic decision rules to evaluate public works projects.
  - The Corps doesn't minimize risk, it maximizes net benefits while considering costs and benefits associated with risk.
  - Comparing $ costs with non-monetary costs and benefits.
  - Risks in costs.
  - Cost-risk analysis and its potential for improved analysis in Corps' decisions.
  - There are a variety of measures for valuing loss of life and they are not consistent.
  - Concepts and discussion of "honorable venture" and "honorable failure."
  - Honorable failure: when is adversity worthwhile or justified?
- *Uncertainty*
  - Uncertainty sources, types, and theories.
  - Classification of types of uncertainties and choice of methods to deal with them.
  - State-of-the-art in quantifying uncertainty.
  - Flood protection design under uncertainty.
  - Think "uncertainty," not risk *per se.*
  - Lack of good reliable data.
- *Extreme events*
  - How to estimate probabilities of occurrence with no historical record (e.g., extreme events).
  - Value of protection against extremes.
- *Infrastructure, security, and other applications*
  - Application to wider range of problems; e.g., dam safety, critical infrastructure.
  - Work on infrastructure security and risk analysis.
  - Dam safety risk-based decisionmaking, risk acceptance, and tolerance of dam safety decisions.
  - Climate change and associated risks.
  - The Corps of Engineers' risk procedures and possible improvements.
  - National Research Council critique (HEC-FDA).

- *Theory*
    - People (organizations) are now applying some of the formal theoretical approaches with success.
    - Development of mathematical formalism for risk assessment.
    - Connections with other aspects of decision sciences.
    - Is risk analysis scientific if experts cannot agree?
    - It is science-based but not science.

2002

- *All-hazards risk management*
    - Identification of the need for all-hazards risk management.
    - The need to address risk analysis and management from an all-hazards perspective.
    - Need for multi-hazard risk-based decisionmaking process.
- *Cyber and control systems*
    - Intrinsic properties of large-scale systems, including redundancy, resilience, robustness.
    - How to undertake cyber-terror with risk management.
    - Vulnerability to cyber-terrorism.
    - Vulnerabilities of control systems.
    - Cyber security.
    - Better understanding of information security (threats and counter-measures, etc.).
    - Vulnerability/risks to control systems.
    - How to protect against attacks on control systems which are designed to be used by lower-level people (in hierarchy).
    - Vulnerability of our control systems and cyber risks.
    - The extreme vulnerability of the cyber-systems to penetration and its critical role in infrastructure protection.
    - *Terrorism*
        - Terrorism and how it is being concluded/studied/modeled.
        - Terrorism raises some separate research questions that require different methods.
        - Terrorism can be addressed systematically in terms of vulnerability, but requires new approaches.
        - Distinction between human terrorism and natural hazard analysis.
        - What we should do to protect our water resources from terrorist attacks.
            - Identify weak spots/likely attacks.
            - How to deter/prevent attacks.
            - How to anticipate attacks.
- *Maths*
    - Probabilistic decisionmaking risk analytic methods will not work for terrorism.
    - The game-theory formulations suggest that the bad guys have a crucial advantage: their ability to select their target, so "hardening is not a good strategy" (Kaplan).
    - Probability not a useful concept when human intent replaces randomness of nature.
    - Fuzzy-set paradigm of risk analysis.
    - Game theory.
    - The concept of probability vs. possibility or prior probability, which I think works great in the ecosystem risk assessment.
    - The inadequacy of existing risk-based probabilistic approach to what is inherently a fuzzy problem.

- *Interconnectedness*
  - Interconnectedness of infrastructure.
  - The need to address interdependencies of critical infrastructures, risk, and vulnerabilities.
  - Take greater account of population at risk in conjunction with infrastructure at risk.
  - Understanding infrastructure interconnections for predicting consequences of willful attacks.
  - Risk due to infrastructure interconnectedness.
- *Other*
  - Definitions of vulnerability.
  - VAG, Inc. VAM model.
  - Homeland security and risk.
  - OHS/DHS thinking.
  - Definitions of terms such as risk (= threat x vulnerability x consequence) and terrorism.
  - Human values and diverse value systems of stakeholders are often the driving decision parameters, rather than the supporting structure analyses.
  - The concept of vulnerability vs. threat vs. risk and how we model these systems.
  - The need for engagement with the private sector in addressing critical infrastructure protection.
  - Vulnerability.
  - The old form of risk assessment cannot work in the new wave of problems that we face (environment, cyberspace, dams, security).
  - Problems of using risk assessment for malevolent risk.
  - Develop better lines of communication between the disciplines/infrastructures (i.e., better understanding of shared terminology).
  - Definitions for vulnerability, risk, and threat—variety of viewpoints.
  - Sensitivity of systems to information assurance component.
  - Variety of models and then use in expressing risk concepts.
  - Take account of resources (such as time, money, people, etc.) in employing measures for hardening systems.
  - How to deal with uncertain probabilities (that is, Hack's presentation).
  - Relative merits of multi-objective decisionmaking vs. multi-attribute expected utility theory.
  - Risk communication and an emphasis on economic tradeoff analysis to communicate risk management
  - No one person, system, group, philosophy has the answer to all problems.
  - Protection of the ecosystem (Everglades) multi-objective risk-cost-benefit approach is needed to see if the plans are feasible and cost-effective.
  - Need not only to protect against terrorism, but also against psychological attacks (i.e., changing water color/taste can cause people to think that there is an attack).
  - "Wear terrorist hat" for identifying plausible terrorism scenarios.
  - In-depth extreme-event analysis, classifying extreme events into three categories: natural, accidents, and willful.
  - 10^{-9} means one second in 35 years.
  - Extension of i-o techniques to terrorists' impact analysis.
  - Terrorist threats and modeling of fault/event trees.
  - Extreme events for dam safety risk assessment.
  - Identifying and classifying exogenous threats and visualizing unique initial events.

2. The most important new ideas/concepts of this conference:

1980

- The use of sensitivity analysis to integrate the uncertainties in risk/benefit analysis.
- The different perceptions of the use of risk/benefit analysis and the lack of thought given to "the process" versus the methodologies.
- Lack of consensus concerning the definitions and procedures that are desirable in risk/benefit analysis.
- Current status of risk/benefit use in water resources (mainly supply), and the importance of political issues in water resources.
- Lack of standardization of risk/benefit analysis, thereby making it difficult for the practitioner to accept.
- Exposure to various approaches and views.
- Stratification of risk/benefit analysis by purpose or other principles.
- A better understanding of the state-of-the-art of implementing risk/benefit analysis and of its complexity.
- A better understanding of the degree of difference in perceived and actual risk.
- The infancy of risk/benefit methods, and the potential for development.
- The imperative need for prominent display of the uncertainty band in reported results.
- Some understanding of:
    - People's perception of risk.
    - Approaches to risk/benefit analysis and recognition that a single approach is not available nor desirable.
    - Probabilities of rare events and difficulties of estimating them.
- The identification of technical leaders in the field.
- The high level of confusion about risk and uncertainty in the context of decision analysis.
- The low level of progress to-date and the rudimentary state-of-the-art of risk/benefit analysis.
- The idea of relative probabilities proposed by Bill Rowe makes sense, but only as a screening procedure to eliminate some of the least-considered outcomes.
- As professionals we still are not willing to discuss openly the risks of structural failure. In some respects we are not able to do so, e.g., foundation failure from undetected weakness.
- The working of Congress; background of the principles and standards.
- Lack of confidence displayed by speakers in risk/benefit analysis today—great deal of "negative thinking."
- North's applications of simple probability distribution function.
- Rowe's conception of risk.
- Slovic's concepts of perception and risk statements.
- Hall's and Haimes' concepts on risk-benefit analysis in a multi-objective framework.
- Campbell's 3 x 3 matrix approach.
- Howard's concept of micromorts.
- Imperative that a consistent terminology and analytical process be adopted.
- Adjusting uncertainty and risk assessments to suit the type and quality of information available.

1985

- Risk-based decisionmaking doesn't replace common-sense approaches.
- The importance of political ramifications in decisions.
- Better able to recognize the limitations of risk analysis.
- There are some modeling alternatives to "expected value" which allow a decisionmaker to deal with low-frequency/high-consequence events in a future period (Haimes' presentation).
- There is some good work on methods for extending data curves from and to remote events (LeMehaute presentation).
- Liability transfer.
- Availability of improved analytic techniques to model uncertainty (partial applicability).
- Is it cheaper (more efficient) to accept failure?
- Presentation of maximal distributions and least favorable in a risk analysis framework.
- Standards are extremely difficult to revise—include an updating procedure.
- Use of risk analysis as another technique in decisionmaking—can be done on basis of insurance.
- Illustrating examples of applications of risk analysis in areas of planning, design, and management. (I would have liked to have seen more of these; several presented were excellent.)
- Risk partitioning as a method.
- Risk utility measurements and decision trees.
- Under extreme conditions of uncertainty, a decisionmaker should avoid the irreversible commitment of resources, but also try to bracket the risk.
- It is difficult to specify single ideas and concepts as more important than others. In my case, the conference allowed me to get an overall feeling for the current thinking and philosophy in the profession.
- Apparent acceptance of risk assessment.
- Divergence between "willingness to pay" and "payoff" (i.e., willingness to accept payment for risk-taking); inconsistencies in people.
- Disagreements with existing standards and willingness to change or adapt.
- Using risk analysis to help decide on a certain alternative in the planning process for a project design, rather than going with a conservative analysis.
- Use risk analysis at the beginning of project planning and not as an afterthought.
- Accounting for future as well as present conditions in risk analysis and decisionmaking in the project design.
- Concept of risk as a multi-objective problem.
- Concept of risk partitioning.
- Idea that tradeoffs should occur among commensurate alternatives.
- Application of risk analysis in breakwater design.
- Application of risk analysis in dam safety analysis.
- Fiering's equipment approach of "certainty equivalence" is a fresh look at an old idea.
- Better understanding of risk analysis concepts.
- The state-of-the-art of risk-based decisionmaking in the academic and agency environments.
- Background and better understanding of the methodology of risk analysis for decisionmaking.

- Lucien Duckstein's talk on combining probabilistic design criteria.
- Bernard LeMehaute's presentation as a form of Bayes' statistics.
- Yacov Haimes' presentation of analyzing different spectrums of the probability distribution function.
- How to think about ways to deal with imprecise values, probabilities, and preferences.
- Liability considerations in professional engineering risk decisionmaking and analyses.
- Gaps in engineering education for decisionmaking.
- Cost of lack of information. Bayesian framework including value of additional information in risk analysis.
- Risk analysis optimization models that lead to a set of alternatives that often don't vary that much in cost (robust solutions).
- No real common agreement about what is risk analysis.
- Engineering standards are not sacrosanct to engineers.
- Several different concepts all converging on the practice of risk analysis (multi-attribute models, performance-reliability indices, risk partitioning).
- Overview of various situations in water resources in which risk analysis needs to be applied.
- A better understanding of the reluctance of engineers to use risk analysis.
- Engineers should design for failure (with proper public notice—e.g., barrier-islands) vs. fail-safe systems. Calculate cost of avoiding failure. Make it explicit.
- Many research models don't improve much over common sense. They help option analysis, however.
- Lack of data truly limits most risk assessments. The value of the reduction of uncertainty is a useful concept. Risk-based decision analysis utilizes data until exhausted, then performs sensitivity analyses where uncertainty exists in order to narrow variables.
- Multi-objective tradeoffs are necessary.
- Risk assessment may increase liability of decisionmakers.
- Value of risk analysis in problem-solving and decisionmaking.
- Exposure to several techniques to perform risk assessment.
- Philosophy of risk, its perception by the public, and the estimation of real risk.
- Difference between risk analysis and uncertainty.
- Vast array of techniques.
- Liability transformations.
- Instability of critical variables.
- The inherent weaknesses of expected-value models.
- The over-reliance on the PMF (probable maximum flood) mentality.
- The real role of risk-based decisionmaking as a "process" tool rather than a "final product" tool.
- Limits of risk analysis methods, and defensibility of the results.
- Decisions should consider the broadest and most complete set of available alternatives.
- Legal, philosophical basis for risk management.
- Risk as one of multiple objectives.
- Existing agency procedures.
- Risk analysis provides a tool through which a decisionmaker can validate and replicate, and thereby justify subjective judgments through a logical procedure.
- Dependent, sequential portfolios and independent portfolios as two separate and distinct categories of the "portfolio problem" within risk analysis.

1989

- Criteria and alternatives in dredging—application of MCDM methods.
- A body of literature on the behavior of decisionmakers in non-engineering fields.
- Use macro models for macro problems and micro models for micro problems. Don't cross these.
- Can expected regret be used as a decision criterion for rare events?
- Don't use expected value for rare events.
- Numbers are much less important than the communication process as uncertainty increases.
- Social scientists and physical scientists can do excellent or sloppy work, but we put on blinders when looking at our own.
- The role of risk estimation in a comprehensive decisionmaking process.
- Extreme-event and Bayesian methods as complements to current methods for risk estimation.
- The nature of the decision can determine the level of resolution in the objective analysis.
- Need to keep multiple-decision criteria for a fully meaningful evaluation.
- Need to compare actual behavior to intended (e.g., survey) behavior.
- Need to learn fuzzy-set methods.
- Truth when communicating risk is best approach to follow.
- Usefulness of fuzzy-set theory.
- Systematic process by Bowles for assessing risk in dam safety decisions.
- In rare events, expected value is not meaningful.
- Fuzzy-set logic may be useful where data is scarce.
- Utility function is not well-correlated with expected value.
- Application of fuzzy-set theory to ill-defined problems.
- Risk partition as an alternative to expected value.
- Risk perception/communication and role of individual public preferences along with technical experts.
- Actual use of fuzzy sets in analysis.
- Multi-objective decision trees.
- Aggregation of component risks.
- Abuse of risk-based decisionmaking.
- Risk assessment of dam safety.
- Overestimation of the impact of climatic change.
- Case studies of risk assessment/risk analysis were useful.
- The importance of process in making risk decisions.
- The potential empirical importance of informed consent.
- The role of models in ecological risk assessment.
- Impossible to validate models.
- Public perception varies with time and the way polls are formulated.
- Models do not always agree with future results, but good decisions are based on them anyway.
- Fuzzy sets.
- Federal agency procedures and formal process for risk assessment
- Risk analysis has led to little confidence in expert opinion.
- Multicriterion decisionmaking.

- Importance of expert judgment.
- Possible use of fuzzy sets.
- Risk vs. recovery in ecosystems.
- Risk perception vs. attitude change in population
- Inconsistency of agencies in dealing with risk.
- Diverse perspectives on global warming.
- Problems with using GCHs to inform spatial decisionmaking.
- Role of the review board in the high-level radioactive waste disposal program.

1991

- The @Risk software.
- How engineers "think."
- Training deficiencies of engineers relative to probability and statistics.
- Practical risk assessment applications (by Neil Parrett).
- Aversion to epistemic uncertainties (by Pate-Cornell).
- Software.
- More flexible interpretation of standards to accommodate the use of risk analysis in design, planning, and management.
- Probability interpretation of risk.
- Use of risk management software.
- Use of uncertainty in decision models.
- Software that is available.
- Wetlands practices, assumptions, and policy gaps.
- Risk-based fuzzy-set theory.
- Utility theory using epistemic and aleatory uncertainties.
- Decomposition of single multipurpose to multi-single-purpose reservoir.
- Modification of utility theory.
- Software is available to perform many of the risk analyses.
- Triangular distribution and variations thereof have been found to have wide application.
- Fuzzy-set theory has been gaining acceptance.
- In the present rapidly-changing environment, manage change or change management.
- Do not attempt to find the perfect solution to the wrong problem.
- Reality, ecology, and economics are full of surprises.
- Software's potential contribution to risk analysis.
- Fuzzy sets are not statistics.
- We can deal with risk concepts in terms of positive expected outcomes or in combinations of outcomes: positive, neutral, negative. We do not need to always deal with "things gone or going wrong."
- Put some risk analysis on criteria used to evaluate priority of dams for budget and study.
- Use risk analyses on testing of emergency warning systems.
- Consider how to evaluate expert input for value/cost distributions.
- Guidelines for evaluating rehabilitation of Corps of Engineers' projects.
- Risk sensitivity models for flood-control analysis.
- Constrained reservoir modeling: a fuzzy-set approach.

- Recognition that models that are used to measure probability have various degrees of uncertainty, and that a choice of a particular model can prematurely discard real ranges of uncertainty without considering them.
- Sensitivity analysis on the model constraints so that the critical driving parameters in a model can be determined early in its development.
- Actual data drives the usefulness of model results, not the structure.
- Use of fuzzy sets on risk analysis.
- Analysis of epistemic uncertainties.
- Risk analysis software developments.
- Pate-Cornell's notion of ambiguity attitudes was very intriguing.
- I was introduced to the application of fuzzy-set theory and found that useful.
- Dudley's water market scheme for Australian reservoirs.
- Threshold is very important in the decisionmaking process.
- Expected-utility decision theory.
- Fuzzy analysis.
- Flood warning threshold.
- Preference structures [of the public and the administrators] that shape water resources management policies.
- Perceptions of economic mechanisms (e.g., discounting of future benefits and elasticity of water demand).
- Risk assessment tools for application should be simple and transparent if there is hope for a widespread adoption.
- Uncertainty reduction benefits can be quantified in a broad sense, including more "efficient" disaggregate behavior.
- Role that underlying value systems can play in risk preferences.
- Opportunity to calibrate teaching and research activities.
- Current state-of-the-art of risk support software activities.
- Meet new people, renew old relationships.
- The potential of using fuzzy-set theory to solve problems in uncertainty.
- I value the potential usefulness of software such as @Risk and AWARE.
- Fuzzy-set formulation of risk.
- Multicriteria framework of infrastructure rehabilitation.
- The environmental (ecological) risk formulation.
- Cultural conflicts between professions is a great stumbling-block to effective decisionmaking.
- Fuzzy-set theory as a new tool.
- The knowledge of others working on similar problems whom I can call on for help and advice.
- Awareness of the new work on risk-based cost/benefit analyses.
- The computer packages available to help with risk analysis.
- The need to treat each wetland as a specific case—the inappropriateness of generalizing.
- Risk management software is most directly useful.
- Suggested risk management procedures placing data and decisions in hands of water users offers intriguing possibilities.

1993

- De-biasing incident databases. (3)
- Advances in decisional-risk software.
- Different methods of ascertaining risk attitudes.
- Four types of uncertainty described by George Apostokalis are a good separation.
- Correcting data biases using expert judgment to eliminate nonsense and capture behavior patterns.
- Identifying critical uncertainties or critical parameters in the myriad of uncertainty sources is a useful shortcut to extensive structuring and computations.
- Alternative method of comparing valuations of risk reduction.
- Do not ask expert for his/her opinions—ask for evidence in all confidence.
- Combining risks and confidences in risk-cost analysis.
- Concept of "de-biasing" data sets may prove useful in effective use of "all" information.
- Use of commercial software packages for decision analysis with some better sense of appropriate applications.
- New ideas for presentation strategies.
- Software availability.
- Advances outside of water resources.
- Can point to COE activities as good precedent for other applications.
- Software is very limited and inefficient; only stand-alone systems are available.
- Many engineering applications that have not been touched can be performed; applications are only in their infancy.
- Application of software tools.
- De-biasing of databases.
- Use of risk/reliability in other fields.
- Len Shabman's work on public perceptions of risk and of avoidance, and comparisons of CV and other methods.
- Liability problems with moving to risk analysis from "accepted" standards with precedents.
- The software displayed.
- Importance of presentation results.
- Data de-biasing.
- Availability of commercial software.
- Increased emphasis on statistics and probability in engineering.
- Description of risk analysis in the Corps' flood damage reduction guidelines (Burnham and Moser).
- Risk analysis in major rehab projects (Leggett).
- Commercial software for risk analysis (Yoe).
- Proliferation of handy risk analysis software.
- Innovative ideas and insights on structuring problems within risk context.
- Look for precedent analyses in electric and gas utility problems.
- Availability of software for decision trees, influence diagrams, and probabilistic project management.
- Bayesian approach to de-biasing an incident database.
- Concepts of risk decision procedures that are being used in making public policy.
- Procedures for de-biasing a database.
- Meaning of probability distributions in risk decisionmaking.

- Reliability vs. economic considerations for decisionmaking.
- Objective vs. subjective de-biasing of databases.
- Available software for risk analysis.
- Discussion on system reliability.
- Knowing who is doing what; broadening the communication network.
- Significant advances are possible in risk assessment applied to a variety of topics.
- Difference between risk analysis and risk management.
- Rowe's range estimates for decisionmakers.
- Use of experts to identify adjustment factors; i.e., Pate-Cornell's method with de-biasing.
- Wide range of off-the-shelf computer packages.
- Software risk session and the information on the availability of software for risk and decisionmaking.
- Availability and extent of software that has become available and should be tried/evaluated.
- Importance of Bayes' theorem for risk-based decisionmaking.
- Techniques being developed for representing confidence limits in combined factors.
- Uncertainty in sedimentation rates.
- US Army Corps of Engineers' sphere of influence/system boundary—be careful of the assumption of it.
- Eliminating change an interesting, significant topic—one of the most important presented.
- Bill Rowe's presentation/topic—perhaps the most important of the conference.
- Water resources planning/management practices and contemporary climate change are adequate to deal with climate uncertainty.

1995

- Dan Willard's ecological principles and importance of scale effects.
- I need to investigate fuzzy-set methods further to judge the relevance to my work.
- The nature of new legislative moves regarding risk analysis.
- Fuzzy arithmetic in risk-based decisionmaking.
- Reliability analyses that deal with time-dependent needs.
- Subjective probability assessment.
- Dam safety risk assessment.
- Risk-based benefit-cost approaches to real problems.
- Fuzzy number theory; fuzzy-rule-based methods.
- Psychological influences on risk perception and management.
- Recognition of the analyst's risk biases.
- The Corps of Engineers has finally gotten into risk-based analysis business procedures in their work. This was evident in the work at the Waterways Experiment Station. This was an excellent application.
- There seems to be a trend to have new engineers make engineering-related decisions based upon very uncertain economics and very uncertain public impact, at the expense of good engineering/scientific reasoning.
- The fuzzy-set ideas may have some future in answering some problems in risk-based decisionmaking.
- Risk and uncertainty analyses do not seem to be applicable to climatic change. The uncertainty is so large that the analyses are meaningless.
- Applications of set theory to imprecise and uncertain risk problems.

- Potential areas of practice and research.
- Research and ideas that are not in the currently existing literature.
- The wide variety of problems, projects, programs, etc. to which methodologies are applicable.
- Len Shabman's distinction between weakly and strongly risk-averse philosophical perspectives for risk assessment/management.
- Reinforced the conclusion that current biological/ecological models have limited predictive validity.
- Extremes and cumulative effects are more important than typical or average cases in design/management.
- The discussion surrounding precision and accuracy.
- The similarity between network flow in groundwater, energy flow in ecosystems, and information theory.
- The widespread belief in the reality of models for this and that, which have not been carefully field-checked.
- Application of fuzzy calculus for risk assessment.
- Subjective vs. objective probabilistic risk analysis.
- Role of extremes in decisionmaking.
- Report on climatic change studies (Pacific northwest, southwest US): demand change overshadows climatic change for now.
- Federal agencies need standardized procedures, but are slow to move.
- Subjective probabilities are used; let's try fuzzy logic combination or opinion.
- Legislative initiatives on risk management.
- Understanding the process of evaluating climate change.
- Understanding the procedure for risk analysis of waste disposal.
- How risk analysis might be used to prioritize research programs in a large organization.
- Possible use of the process in allocating funds among competing approaches to a common goal.
- Improved ability to evaluate research proposals—i.e., better appreciation for topic and its current applications.
- Fuzzy-rule-based methods.
- The limitations of standards-based design.
- Risk analysis addressing regulation.
- Expert opinion.
- Fuzzy-number applications.
- Views of other disciplines.
- Those making the decision are more important than the analysis supporting the decision.
- All programs must show some benefit beyond satisfying intellectual curiosity.
- Fuzzy-based rule making.
- Defining ecosystem value/health based on adaptability.
- Partitioned multi-objective risk assessment for extreme events
- Statistics of extremes.
- Importance of scale for ecological health.
- Sandia model for waste-management decisionmaking.
- Applying the percolation network approach to simulating geologic scale multi-phase transport.
- Development of fuzzy rules.

- New models for assessment (especially since available in the public domain): Sandia National Lab.
- Stormwater issues: Colorado very helpful.
- Ways to elicit subjective probability assessments.
- Probabilistic model development in risk-based decisionmaking.
- Using risk information in decisionmaking.
- Decisionmaking with uncertainty in data.
- Communicating relative risk.
- Legislative issues on risk.
- Use of stochastic methods.
- Risk associated with groundwater.
- Using BCR in engineering design.
- Can we show that BCR reduces regulatory hassles?
- Significant disparity among disciplines regarding basic definitions of risk, hazard, etc.
- Risk and environmental decisionmaking concepts have been explored and evaluated.
- Communicating in risk analysis is a volatile subject (i.e., how?, who?).
- The cost of risk assessment, cost-benefit modeling, etc., may not be as high as initially perceived.
- The multi-perspective, extended length, close interaction form of this conference would be useful for some problem-solving situations—e.g., theory development.
- The practical method of subjective probability estimation.
-

1997

- Extreme-event modeling.
- Fuzzy-rule-based modeling.
- Uncertainty in health risk analysis.
- Cost-benefit analysis.
- The level of uncertainty really impacts on the level of analysis that is conducted.
- Expert judgment is much more difficult to handle than one recently thought.
- Maximum entropy/probability bounds approach.
- Regulatory approaches.
- Need to characterize uncertainty.
- Use of multivariate analysis in risk prioritization.
- Strategies for managing uncertainty.
- The rehabilitation of structures is a gigantic multi-objective dynamic problem under uncertainty.
- The methodological field must be enlarged not only to statistical and Bayesian methods, but to fuzzy sets, rough sets, belief functions, and possibility theory, among others.
- A few tutorials may be in order—but not at too low a level, please!
- TRIZ.
- Rowe's hierarchy of uncertainty—good taxonomy.
- Classifying robust techniques for uncertainty measurement.
- Interaction of connecting water barrier and levees in Netherlands.
- Systematic planning of navigation, locks and dams, dredging.
- Multi-objective analysis.
- Extreme-value analysis.
- Keep the "experts" away from your "decisionmakers."

- Degree of belief is a key element in risk assessments.
- Need to focus on gathering useful information to support decisions.
- Status of dam safety standards in US and Canada.
- The Dutch approach to issue of sea defense.
- Update on perspectives of planning, evaluation, and risk for water resources at the US federal level.
- Criteria-based ranking for environmental investments.
- Extreme precipitation is related to both ENSO and circulation patterns.
- "Uncertainty" vs. "risk" vs. "surety."
- Dispersive Monte Carlo sampling.
- Wicked decisions.
- Competent error vs. negligent error.
- How to deal with risks and uncertainties, with case studies.
- Dealing with expert opinion and its role in risk analysis.
- Similarity of risk analysis issues across different sources (areas) of decisionmaking.
- QA for Monte Carlo.
- Risk and uncertainty management strategies.
- Definition of terms.
- Uncertainty relative to risk.
- Risk assessment vs. risk management.
- Methods for determining acceptable risk.
- "Probability bounds" methods may be helpful to generate input distributions when you don't know much.
- Use of evacuation models.
- Risk management of portfolios.
- New means for combining distributions.
- Range estimates of parameters and results are required rather than point estimates.
- New approaches to dealing with uncertainty.
- Use of fuzzy sets in climate and weather research.

2000

- *Understanding risk*
  - How I need to define risk.
  - Risk can be reduced by education and/or structural and/or legislative actions.
  - Risks are both real and perceived.
  - Notion that risk acceptance is not constant.
  - Concept of risk reduction influencing increased risky activity, substituting certain benefits for safety.
- *Risk communication*
  - To be most effective, risk analysis should be decision-driven, i.e., conducted in a way that will facilitate communication.
  - Stan Kaplan's communication theorems.
- *Value/economic aspect of risk*
  - Engineering reliability and economics.
  - Risk-based decisionmaking on Corps' cost estimates.
  - Benefit-cost analysis with risk in both components.
  - Applications that address both reliability and risk jointly in a cost-effective paradigm.

- It is not always risk that drives public perception of the acceptability of engineering projects.
- *Extreme events*
  - Extreme-event risk management associated with water resource projects.
  - Assessment of extreme floods.
- *Application*
  - Protecting critical infrastructures.
  - The dam safety risk-based decisionmaking in navigation.
  - Applying risk management to protect people and buildings from bombing attacks.
  - Application to facility maintenance and repair decisionmaking.
  - The ALARP approach to risk management.
  - Exposure to risk analysis in a wider range of applications and appropriate vocabulary.
- *Theory*
  - Difference between risk management and risk assessment; clear definitions of risk analysis, assessment, and management.
  - Difference between natural variation and knowledge uncertainty.
  - Idea of surety = safety + security + reliability.
  - Risk management techniques.
  - System concepts for risk—thinking about what we do not know.
- *Others*
  - Blind ignorance vs. conscious ignorance.
  - Humble failure.
  - Opportunity to talk and meet with peers.

2002

- Shacklean theory of surprise.
- Merits of artificial vs. natural control systems and why it matters.
- Intrinsic properties of large-scale systems, including redundancy, resilience, robustness.
- Decisionmaking under risk when you don't know enough.
- Internet and cyber security issues.
- Possibility "theory" and the need to find out more about this to see if it is useful.
- Terrorism raises a number of separate research questions that require different methods.
- Probabilistic decisionmaking risk analytic methods will not work for terrorism.
- Game-theory formulations suggest that the bad guys have a crucial advantage: their ability to select their target. So "hardening is not a good strategy" (Kaplan).
- Application of game theory to terrorist risk.
- Decisionmaking at community level is based on multi-hazard (natural and malevolent) risks.
- Risk communication—trying to explain the risk to others probably looks like the wrong way. The value systems of the players have more influence on the decision outcomes than the technical risk analysis.
- The human factor in terrorism dominates the frequency and choice of attack, not the nature of it.
- Federal agencies are mobilizing in a rational manner to achieve coordinates and proper focusing of resources cooperatively.
- Develop method to deal with cases of limited data.

- Take account of public's vulnerability (i.e., in an acceptable risk context).
- Sharper distinction between prescriptive and descriptive models of risk-based decisionmaking.
- Possibility and fuzzy arithmetic.
- Many useful references:
  - Intruder behavior model, CERT website.
  - Combating terror of terrorism, SCI American.
  - Simulation model, Defense Information Systems Agency.
- Bill Rowe's model.
- Contact w/John Saunders (info security expert) and others knowledgeable in that area.
- Organization and purpose of office of homeland security.
- Interconnectedness of risk assessment and management methodologies across varying disciplines to improve one's own methods.
- Need for calculus/mathematics of possibility, not only probability.
- No matter how expensive the system, or how much security you put in, the terrorist needs only to attack one place to cause problems. System is only as strong as its weakest link.
- Application of game theory to risk analysis (more realistic model assumptions are needed).
- Motivation to study other risk methods based on:
  - QRA and Bayesian.
  - Fuzzy logic.
  - Extreme-event analysis.
- Reality of terrorism and relaxation of simplifying assumptions in risk models (linear, static, etc.).
- Risk communication and visual models can be used in effectively communicating risk issues to stakeholders.
- Game-theoretical approach.
- Fuzzy-set risk analysis.
- Wetland restoration planning.
- Cyber attacks on control systems.
- The emerging response to large-scale disasters.
- Fuzzy reliability.
- Natural hazards may not be helpful in assessing threats.
- Vulnerability is important.
- Need more information to relate vulnerability to consequences.
- Application of game theory to risk management.
- Learning about work being done in government agencies relating to risks of terrorism and homeland security.
- Methods for vulnerability assessment.
- In the areas of risk assessment and management for ecological purposes, the concept of probability vs. possibility and the development of techniques similar to this approach would help us in the management of species, populations, and even habitats.
- Stan Kaplan's thoughts on terrorism and initiating events.
- I have learned new ways of looking at a system that in my deterministic mind were not explored before. Have to add more interconnection in my systems.
- Prioritizing risk.
- Vulnerability of broader connections.
- Risk framework for terrorism.

- The role of state variables in model construction.
- HHM.
- Vulnerability assessment for terrorism.
- Game theory and possibility theory approaches to these classes of problems.
- The answers to three questions that come out of academia are often too academic.

3. The most important issues needing further study in risk-based decisionmaking in water resources:

1980

- Consensus on definitions, uses, and processes connected with successful application of risk/benefit analysis.
- Identification and discussion of successful risk/benefit analysis applied to real-world decisions.
- Rare events.
- Contamination of drinking water—mainly groundwater sources (aquifers).
- Better measures of benefits.
- Methodologies for estimating risks reliably.
- "Acceptable" level of risks for different types and scopes of projects.
- Drawing in the public in order to develop confidence in the results.
- Need for work on probability of environmental impacts.
- Methodologies which reveal social preferences as they pertain to risk-taking.
- Understanding long-term risks that are not catastrophic rare results—e.g., long-term elimination of pristine environments in urban areas, etc.
- How can we educate users, i.e., planners and managers, to think in terms of probable outcome?
- Identification of critical issues in risk/benefit.
- How can we assess, accurately, the expected demands for water resources outputs?
- Methodologies in risk/benefit analysis.
- Dealing with uncertainties.
- Comparative study of risk/benefit methodologies, particularly in connection with the formulation and analysis of water resources projects.
- Methodologies for presenting "reliability" information in project reports, including simplified methods of display.
- Extension and refinement of multivariable decisionmaking analysis procedures for use in the field of water resources where the problems are to be analyzed in a multi-objective framework.
- Develop and test educational material to help people understand risk tradeoffs and actual risks.
- Develop more flexible planning methods to cope with uncertainty.
- Risk/benefit analysis doesn't replace decisionmaking.
- Improve coordination of the entire decisionmaking process.
- Need for post-"project" or post-planning evaluations of the contributions, impacts, and costs of risk/benefit analysis.
- Need to identify evaluation methods for the assessment and presentation of risk/benefit analysis in a timely fashion.

1985

- Incorporation [into risk analysis] of the consideration of the value of human life in terms of dollars.
- Establishment of acceptable and applicable procedures [for risk analysis].
- Methods to use regional data on extreme events (e.g., precipitation and/or floods) for extending site-specific frequency distributions.
- Methods to identify areas where additional data gathering might be beneficial, and ways to estimate potential benefits from those efforts.
- A framework that interrelates, for political priority setting [of risk management options], an array of competing federal programs.
- Several sets of optional assumptions would be required.
- Work to improve data bases as input to [risk] analysis models.
- A process for predicting recurrence intervals of large flood events derived from rare-event occurrence records.
- Role of hazard management.
- Costs and affordability of mixed-strategy risk-management alternatives.
- "Quick and dirty" low-cost techniques, especially to deal with "non-federal" risk problems.
- Standardize risk analysis, as well as conventional (standard) procedures.
- Continue dialogue with sociologists, lawyers, politicians on the problem of risk analysis.
- Education of young engineers and continuing education of practicing ones needs further strengthening. The cost of risk-based analysis is another issue [that needs resolution]. The liability question is not perceived correctly—it may be overemphasized.
- Financial risk of water resources investments which are designed according to fixed standards (emphasis on future vendibility of project outputs via revenue bond financing).
- Specific applications of the various methods available, to determine the most appropriate methods under certain risk circumstances for various design tasks.
- Methods to incorporate risk analysis into design codes and encourage wide use by both the public and private sectors.
- The effect of forecasting alternative futures over the period of analysis on the risk analysis outcome (contribution to risk of forecasting uncertainty).
- More study needed on structural failure modes other than hydrologic causes of failure.
- Liability [issues connected with risk analysis].
- Probabilities of rare events (e.g., PMF).
- How to get public acceptance of risk assessment—e.g., use case studies which demonstrate cost savings [of project design based on risk analysis].
- How to incorporate public values in risk preference modeling (e.g., through workshops).
- The designer's liability in using risk analysis; the pressure to save on design costs—and some laws to protect or enable the designer to use liability insurance.
- Standardization of risk-benefit analysis to real-world situations and [within] acceptable levels of risk.
- Educating people to use and accept risk-based analysis in project designs and deal with its uncertainties.
- [Need for clearer justification of the] relevance of risk analysis to decisionmaking.
- Use of risk analysis where probabilities are poorly defined.
- Additional work on defining terminology; Duckstein's paper is a good starting point.

- Further exposition on how engineer confronts the question of risk: a) accept existing standards and avoid risk of unacceptable design; b) invest resources to prove that that standard should be modified; or c) perform complete risk analysis of system and its components.
- Need to develop a credible database on [structural failure] damages including the morbidity, mortality aspects.
- Simplifying the concepts and computations [required of risk analysis] so that journeymen engineers can apply concepts.
- Develop [risk analysis] methods, process, procedures to present information so that "the decisionmaker" can understand it.
- The probability assignment for rare events.
- Assigning some values to life or developing a method to incorporate [these values] into risk analysis.
- Use of methods for risk analysis in actual (real) projects from various participants—these case studies should represent various types of projects and situations.
- The replacement of standards with a more rational [evaluation] approach (it is not easy to change the established practices).
- Yacov Haimes suggested an excellent idea for a future meeting where several risk analysis methods could be tested and compared for common case studies.
- How to apply [risk analysis] techniques.
- What the techniques actually do.
- What the techniques cannot yet accomplish.
- Cost of standards at various levels of the probable maximum flood (PMF).
- Cost of various safety factors (used in engineering design).
- Regret analysis leading to robust decisions.
- Measurement errors in components of risk assessment: how important?
- Acceptance of economics of risk assessment in engineering analysis and design.
- Risk assessment in other water resources planning topics. Level of protection in levees, waste disposal, dredged material disposal.
- How to incorporate loss-of-life considerations into risk analysis.
- How to estimate "willingness to pay" for risk reduction.
- How risk analysis can help reassess engineering design and environmental standards.
- Develop a set of common definitions for risk analysis: reliability, safety, risk, uncertainty, etc.
- A risk analysis approach to determining a standard: i.e., use risk analysis in the procedure which comes up with the standard (e.g., PMF).
- More work on alternatives to "expected value" for decisionmaking.
- The possibility of developing more uniform, formal (but not necessarily analytic) approaches to risk analysis.
- Real-world risk assessments with [implementation] results would be helpful—did risk analysis work? Contrast successes vs. failures of risk analysis approaches.
- Human health risk issues (i.e., toxic chemicals).
- Need for standard method. Most engineers do not choose among hundreds of different methods. There is a growing trend for new requirements and assessments (e.g., environment impact statements) to become a part of a process but [they] are not seriously used in the decisions. As a consequence, the engineering profession is becoming like the "fireman" on a diesel train. Let's reduce "mindless" analysis.
- Data development on probabilities of system failures and their modes of failure.

- Need to define risk analysis.
- Need to determine under what situations risk analysis is the preferred methodology, so that its application is not second-guessed even under recognized uncertainty.
- Additional research in estimating the probabilities or probability distribution of extreme events. High risk is associated with events that are poorly defined statistically.
- Quantification of environmental damages is important for risk analysis of decisions that severely affect the environment as well as for tradeoff analysis.
- How risk analysis techniques are applicable to real-world problems.
- Risk perception.
- Development of a more consistent decision calculus.
- Uniqueness of risk assessment compared to all other methodological strategies and tactics.
- How to deal with very-low-probability events.
- Distinction between estimation of risk vs. management of risk (how to decide what level of risk should be accepted).
- What types of alternative models are there in addition to the "traditional" expected-value models?
- How to incorporate life safety (or mortality) into the decision process.
- Fuller and more extensive use of available historic data.
- Preference ordering.
- Developing procedures for assessing a modification to a channel system for flood protection. Such channel systems have multiple subjective uncertainties and multiple constraints which can be placed upon their design.
- The most appropriate situations in which to use risk analysis (e.g., planning, design, operation).
- Analyst-client communication.
- Probability of extremes.
- Sensitivity of decisions to probability considerations.

1989

- How to accommodate public perception of risk in locating hazardous waste or nuclear sites.
- How to reconcile probabilistic risk issues with development of regulations.
- How can we influence decisionmakers to consider risks rationally before determining the final allocation of resources to address a perceived problem?
- Effective risk communication with decisionmakers and the public.
- Case studies where risk assessment has been used and its usefulness/input on decisions.
- How to effectively manage and use expert opinion.
- Would have been nice to have included (1) legal perspectives and/or issues related to risk assessment, (2) more case studies, and (3) some kind of "objective" assessment of the state of the art in risk assessment in our fields and others.
- Multicriterion decisionmaking.
- Importance of expert judgment.
- Possible use of fuzzy sets.
- How to integrate the public's perspectives and expectations into risk-based decisionmaking.

- Dose impacts.
- Empirical testing of ecological models.
- Incorporating expert judgment and opinion into low-probability/high-consequence events.
- Developing separate uniform risk analysis protocols for health- and safety-related technologies.
- NAB/NRC study/overview of conceptual approaches to risk analysis/reliability analysis.
- Abuse of risk decisionmaking.
- Risk assessment of dam safety.
- Overestimation of the impact of climate change.
- Communication of risks/benefits.
- Use of risk analysis to mediate conflicts and minimize social costs.
- The importance of the process of making risk decisions.
- The potential empirical importance of informed consent.
- The role of models in ecological risk assessment.
- Risk assessment techniques.
- The use of "fuzzy" sets rather than probabilistic descriptions.
- Effective representation of relative levels (catastrophes vs. voluntary risks).
- Connecting social science and engineering in calculation and description of risk.
- Are all risk analysts this cynical?
- Public communication of risk issues.
- Simplifying models for policymakers' understanding.
- Can public/governmental decisions be made which include substantial risk?
- When do the studies stop and the work begin?
- Flood warning and evacuation system.
- Risk of extreme events.
- Fuzzy sets and risk analysts.
- Can expert judgment be used as a substitute for data, and how?
- Effective decisionmaking with situations involving rare events.
- Where is risk analysis being abused or misused, and what are the costs of such abuse?
- Processes for the generation of scenarios, alternative designs.
- I, too, probably would have liked to see a little bit more of impact analysis included in the discussion of problems and decisionmaking.
- The implementation of methods and tools/processes in computer-based environments. Decision support system (DSS): not a large measure, but perhaps one or two papers on the subject would help round out the content of the conference.
- Sensitivity of polls' results on the types of questions.
- Risk management under uncertainty.
- Interpretation of quantitative risk analysis results.
- Relating methods to the nature of the decision.
- Identification of relevant model structure and parameter estimation for different decisions under uncertainty.
- Better cross-discipline fertilization of perspectives, methods, etc.
- How does EPA decide on remedial actions (criteria, alternatives)?
- Use of models for regulation of toxic substances.
- Uncertainties in risk analysis for extreme events (floods, other catastrophic events).
- Incorporating risk aversion in a sensible manner.

- Integrated assessment between teams of social and environmental scientists for problems which have impacts and feedbacks in both directions: physical environment ←→ human behavior.
- Need for studies of the roles of scientists in particular environmental problems.
- The need to develop flexible decision rules (and methodologies for determining these) which allow for adoption and uncertainty in both the physical environment and human preference factors.
- The conflict between need for simple national regulations and detailed site-specific science.

1991

- Methods of training existing engineers in probability and risk analysis.
- Comparison of fuzzy results to probabilistic results.
- Flood control (combined structural and nonstructural measures).
- Extreme events.
- Risk management and use of output from risk for decisionmaking by non-technical managers.
- Methods and approaches to bring state of practice closer to state-of-the-art.
- Relationship of proposed risk assessment methods to current public budgetary decisions process.
- Relationship between new technologies and risk (example: use of expert systems for risk management).
- Quantification of risk.
- Use of risk as one objective in water resources management.
- Differences between immediate catastrophic risks (e.g., dam failure) and long-term cumulative risks (e.g., water).
- How to encourage the use of risk-based decisionmaking regarding long-term environmental risks.
- Quality of data needed for the models.
- Risk analysis for floods is extensive, but it is relatively insufficient for droughts.
- Risk-based decisionmaking in conjunctive use of groundwater.
- Software/risk intersection.
- What is risk average? Expected value analysis.
- Validation of models.
- Methodologies to secure better impact on models from decisionmakers.
- Translation of sophisticated techniques to formats that are useful to average engineers/economists.
- Operation of water resources systems with multiple decisionmakers.
- Sustainability of irrigated agriculture.
- Design and operation of regional water resources systems in anticipation of a possible climatic change.
- How to bring cost sensitivity to design of projects.
- How to open up projects (infrastructures) for reevaluating their justification before rehabilitation is done.
- Whether very-long-run uncertainties, for example, warrant as much professional effort as is now being directed at these problems.

- Resolve issues set forth in long-run uncertainties.
- Integration of engineers' ideas of fuzzy-set theory with economists' ideas of expert decisionmaking.
- How to understand the practical application of risk analysis to identify more appropriate "planned outcomes."
- Model for comparing changing values on water use and allocations or utilization of available supplies.
- Continue studies on social costs of environmental policies and program implementation alternatives.
- Better emergency management response to protect lives, properties, and project facilities.
- Integrated systems risk analysis.
- Bayes' approach and its application as compared to fuzzy-set approach.
- How do we test the validity of expert value judgment?
- Does data drive the structure in risk models, or vice versa?
- How do we better understand the uncertainty in the models used to measure probability?
- Associate engineering modeling and simulation with risk analysis.
- Global change in risk analysis.
- Remediation of contaminated sites.
- Education: we need to address how we are training risk analysts.
- Now that we have been working this specialty for some time, is it mature enough to integrate it with the work of the classical risk management (i.e., insurance) and health and environmental work in this area? I believe it is time to do this.
- To bastardize the term *technology transfer*—pushing the frontiers is essential and must continue but we need to pass along our knowledge to practitioners as well as to graduate students. We need to make this framework and some of our models more accessible.
- Decision support systems for drought and flood management are needed (practical software form).
- Risk associated with using fertilizers and pesticides on groundwater resources.
- Using GIS and expert systems to determine the risk associated with the modeling.
- Using fuzzy analysis.
- Safety goals in risk management.
- Economics of water resources risk management (i.e., risk of shortage, risk of failure, costs of human safety, etc.).
- Need to design and formulate situations or experiments where actual behavior can be used to infer risk preferences, rather than expressed preference through interviews or questionnaires.
- Systematic study of the impact of computer-aided tools on adoption of risk-based approaches.
- The roles of the risk brokerage industries—legal and insurance.
- Fuzzy set as related to risk analysis.
- The value of information in risk-based decisionmaking.
- The impact of the discount rate in modeling and decisionmaking.
- Cost-effective acquisition of data to support risk analysis.
- Assessment of B/C of risk assessment and management.
- Technology transfer between risk-based domains such as health, insurance, lotteries, stock market.
- The establishment of a conceptual framework within which risk-based decisionmaking can be performed for diverse problems.

- Developing user-friendly software which can be used as "tools" in practice to apply these concepts.
- Determining the value of expert opinion.
- Climate-change-related risk.
- Risk management under uncertainty.
- Ecological risk management.
- Definitions.
- Applications: what decisions were made based on an analysis.
- Cross-professional communications.
- The need for more work on risk-to-whom—i.e., a sharper focus on just who is bearing the risk, and whether they are the most appropriate persons or agencies.
- Update of work-in-progress (and similar new work) presented here.
- Incorporation of information risk with climatic-based uncertainty for highly variable climates.
- Without trying to impose uniformity of thought and approach, the definition of terms and concepts must be made more uniform if public decisions are to be positively influenced.
- Concepts of risk assessment and risk management need to be communicated to the public and its leaders clearly, simply, and without raising levels of alarm.

1993

- Uncertainty in the states of nature; i.e., how to capture uncertainty in states of nature in parameter estimates used in models of natural systems.
- Specific pdfs with limited or no data.
- Relationship between model and parameter uncertainty and risk-based decisions.
- Cross-agency uniformity—FCCSET.
- Legal constraints.
- Training the public to think in terms of probability.
- Training the public to understand that there is no certainty.
- Training the public that this has tremendous implications for expenditures on hazard mitigation and government aid and will have implications regarding our legal process (litigation).
- Communication techniques for risk.
- Incorporation of risk-based techniques in decision processes.
- Data needs and data availability to support techniques (plus cost of getting data).
- Examination of the potential role of markets for rights to water resources, where risk is important; i.e., do we want to move from "here" (very limited reliance on markets) to "there" (much greater reliance on markets for water resource allocation).
- Examination of the practical difficulties of moving from "here" to "there," especially legal constraints and how they may be overcome.
- Examination of the lack of social costs and benefits of moving from "here" to "there."
- Still no accepted evaluation principles for risk analysis and its role in decisionmaking.
- Decision rules and modeling structures are still *ad hoc*—no agreement among practitioners and agencies.
- Risk communication and public decisionmaking still a major problem—experts vs. public preferences.

- Are our "risks" (probabilities) indifferent between epistemic and aleatory (or uncertain and variable) sources of variance?
- What uncertainties are critical to Corps' decisionmaking? Or, what organizational levels and functions does risk best address?
- Is the "how, where, what?" of risk-based decisionmaking adding value to our society? Explicitly, today's reality and the ideal.
- Methods for incorporating risk-based decisionmaking into sophisticated computer models.
- Transfer of risk techniques used in other fields to problems in water resources.
- Standardization of terms and concepts in risk analysis to improve interchange of information.
- Risk management: much information on risk analysis, but need more ideas on decisionmaking with risk and uncertainty information.
- Defining and characterizing residual risk, i.e., "with project" condition.
- Value of risk reduction in non-monetary terms, i.e., ecosystems, safety, and health.
- Social impacts/response.
- Derivation of pdfs.
- Alternatives to Monte Carlo analysis.
- Risk/reliability aspects of groundwater modeling.
- Training on the use of pdfs.
- Methods for handling time preference.
- Better-developed methods to incorporate uncertainty into risk-based design procedure.
- Development of design standards.
- Develop methodologies to interlace hydraulic, hydrologic, structural, and economic uncertainties into the risk-based procedures.
- Impact of public attitudes on policy options and choices.
- Overcoming organizational/institutional constraints on using risk-based methods.
- Improving data validity and reliability for risk-related phenomena.
- Means to identify rapidly critical decision parameters and uncertainty contributors.
- Means to validate expert judgment used to capture behavior patterns.
- Using Monte Carlo simulations to explore uncertainty may be appropriate, but has little to do with risk analysis itself.
- Methods for sorting through uncertainties to provide focus for decisionmaking.
- Need to develop risk communication approaches.
- Issue of economic inequalities; finding ways to expand decisionmaking beyond purely economic constraints.
- Development of tools and methods for effective communication of risk-based decisions to decisionmakers and public.
- How to use risk analysis to build consensus around decisions on risk management.
- How to share liability for adverse (ex-post) outcomes that occur when sound risk management decisions were made in advance (ex-ante).
- Accounting for human factors in making risk decisions; i.e., survey data, etc.
- Verification of models. Can we or can we not validate a model?
- Expertise input to risk decisions: how much influence does it have on the outcome?
- More complete integration of probabilistic approaches in all aspects of engineering.
- Continue unified approaches to the problem.
- More systematic development of databases, e.g., national database on dams.

- Need to understand the dominant impact of organizational behavior on risk-based decisionmaking.
- More research on the reliability of water distribution systems.
- Develop theoretical foundations that relate Bayesian and fuzzy-set thinking.
- Combine existing knowledge (data) with probabilistic modeling for risk analysis.
- Risk analysis of extremes.
- Environmental risk analysis.
- Hands-on workshop sessions to go through case-study applications of risk.
- Continue theory presentations, but mix in more application.
- Look at health/ecological risk in both surface and groundwater issues.
- What role does pollution prevention play?
- There should be a more subjective analysis of global climate change; we obviously can disagree on the validity of the models and resulting probabilities, but we should be able to agree on a methodology and let the alternative data sets drive the results.
- How to go from global climate change to regional foci.
- Need to spend much more effort in looking at ecosystem/water quality impacts of climate change.
- Need for scenario modeling to test whether and under what conditions climate change will lead to a different decision.
- What are incremental and irreversible decisions in long-range water planning?
- Is water distribution system reliability an important issue or not? (Males "nay," Heaney "yea").

1995

- Greater focus on management to optimize the costs or benefits from complete probability distributions of hydrologic events to improve or strengthen financial positions to withstand the impact of extreme events.
- Advantages and disadvantages of community involvement in planning and operating water resource systems.
- Use of fuzzy arithmetic in risk-based decisionmaking.
- How to reduce lead and mercury in Michigan.
- Introduction of more fuzzy-set theory into water resources.
- Need to better define the extreme events.
- Need to show more practical, i.e., engineering-level, applications of risk-based decisionmaking.
- Usefulness of fuzzy numbers.
- How to deal with data quantity/quality problems.
- How to deal with the analyst's risk biases.
- There needs to be an emphasis on using risk-based decisionmaking in conjunction with the engineering of the problem, not to replace it. The physics of the problem cannot be replaced by probabilities and statistics.
- Engineering procedures (risk-based) need to be standardized.
- Extension of the risk-based procedure to appropriately account for the economics needs a major emphasis
- Developing risk-analytic methods that are better linked to the "physics" of the problems studied.

- The profession (Society for Risk Analysis?) needs to develop more uniform/standardized handbooks or manuals, somewhat akin to the ASCE, to develop an industry standard in terms of risk/cost analysis; otherwise, we have chaos.
- Risk analysis normally fits within a broader, multi-objective, and multi-procedural evaluation framework. Risk analysis is often ill-suited and poorly designed for compatibility with the governing framework.
- Better methods of evaluating costs and benefits for problems involving environment and ecology.
- Institutional responses.
- Continue studies on how to incorporate methodologies into decisionmaking.
- Methods for prioritizing water infrastructure rehab projects that balance technical, economic, and political considerations.
- Ways or opportunities to adjust and/or circumvent regulatory constraints to allow for more cooperation between agencies and other stakeholders in planning [implementation of water projects].
- Sustainable ecosystem management—the short- vs. long-term benefits to public health, welfare, and safety.
- Some integration of water quantity and water quality issues.
- Integration over time for risk reduction, public benefit, and environmental cost in a timely way.
- Fuzzy vs. probabilistic risk analysis.
- Developing an integrated risk analysis framework for engineers, ecologists, lawyers, economists, and public administrators.
- Ecological and human health risk assessment and management.
- Communicating uncertainty.
- Have engineers and social scientists harmonize language, concepts, methods, and techniques.
- Risk analysis in ecology, groundwater, and nuclear waste sites is inadequate.
- How to sell risk-based decisionmaking to the public.
- Resolving conflicts among competing interests.
- Educating future decisionmakers on risk analysis.
- Defining data needs for applying the process to natural systems.
- Multi-disciplinary effort to improve communication of risk.
- Risk evaluation of natural ecosystems.
- Education: training individuals in risk-benefit technique for pollution prevention, flood-damage reduction, and accident-frequency reduction.
- Need large-scale, multi-objective case studies to illustrate power/limitations of the process; i.e., how does one apply risk analysis in a river basin where priorities change as a function of position within the basin (reservoirs, in-stream flows, coastal ecosystems)?
- Ways to promote application.
- Developing uniform general techniques of risk analysis.
- Coordination among federal agencies.
- The role of values or world views in the process of consensus-making and mood development.
- A decisionmaking architecture must be made explicit to all. (Many of the presentations examined the question at hand in isolation. Thus, the data from the study was not compared to other national or local environmental needs.)

- Emerging role of sensor technologies with information technologies in water resource infrastructure maintenance, restoration, and/or planning.
- The role of market mechanisms in reducing water quality risks. Bring in Paul Partrey.
- Standard, accepted methods for treating uncertainty.
- The drawbacks and dangers of standardizing risk analysis for the sake of easy acceptance.
- Risk communication and education of decisionmakers, and changing regulations to become risk-based. Also, investigating how the risk assessment process will need to be revised to handle the mandated cost/risk benefit analysis. Specifically, using risk/cost benefits as efficient screening tools.
- Identifying legal/regulatory constraints on the use of risk-based methods—and strategies for reform.
- Communicating to elected officials and the public that risk-based methods are not simply excuses for environmental degradation.
- Defining ecological restoration.
- A holistic approach to subsurface contamination.
- Integrated framework for risk analysis.
- Extreme events.
- Groundwater contamination from agriculture.
- Are we asking the right questions in risk assessment for risk managers? Effects of changes in regulatory environment on risk assessment and management.
- Development of better software tools.
- Ways of dealing with time aspects.
- Extreme-event calculations.
- A more unified approach, with engineering design using reliability-based methods.
- More focus on adaptive managers using real-time control.
- Need more work on comparative risk analysis.
- Need some discussion about the variability needed in local framing conventions and the need for standardization as one moves up to the national level in decisionmaking; i.e., moves from fuzzy to blurry.
- Need information about risk management strategies and how you use results of studies.
- More research is needed in conceptual framing, assumptions, etc. Psychological influences on defining problems, identifying alternatives, and estimating and reacting to risks inherent in water resource problem-solving.
- Applying information-processing technology to water resource management/problem-solving.
- Prioritizing water resource efforts under personnel, financial, time constraints, and others.

1997

- Fuzzy modeling.
- Handling the public perception of risk.
- How analysis is acquired and used in practice.
- Presenting uncertainty to the decisionmakers.
- Bayesian methods in cost-benefit analysis.
- Selecting distributions for extreme events.
- Defining decisionmaking.

- How to be right rather than correct.
- How to better choose/revise measurement endpoints for risk analysis and risk management.
- Modeling purposeful attacks on water systems.
- The role of risk analysis in a total system analysis.
- How government/industry are setting acceptable risk levels.
- Relevance of Bayesian theory to decisionmakers.
- Uncertainty analyses.
- Dealing with long-term, cumulative, uncertain risks.
- Regulatory reform.
- How to fund risk-based decisionmaking in public agencies.
- Regulatory reform.
- Role of risk analysis as part of multi-objective decisionmaking, not vice-versa.
- Education/training for field analysts: top-10 books/software needed; what are colleges/industry/government doing?
- Extreme-event analysis.
- Fuzzy-set workshop (go through application).
- Dam safety/sea level rise.
- How to convey results of risk-based decisionmaking analyses to the decisionmaker.
- Agreement on terminology.
- Dam safety guidelines.
- Multiple failure modes.
- Regulating problems.
- Society's willingness to pay to mitigate risks.
- Comparison of subjective probability vs. quantitative.
- Risk and cost of misspecified models (wrong model).
- Need to establish benchmarks vs. aversion to benchmarks.
- General agreement among the members on the framework for risk analysis.
- To understand how each organization's standard compares to the benchmark.
- To consider risk as just one element of decisionmaking.
- Providing the use of good science in practice.
- Encoding uncertainties in risk analysis.
- How are tradeoffs made reliably?
- Use of good systems engineering to define risk problems; then assess and manage them.
- Bridge the gap between "soft" and "hard" versions of risk analysis (polity and social issues vs. engineering/math models)—significant, possible, probable.
- Provide predictive studies with validation and honest constructive criticism of past mistakes.
- Evaluation and perception of risk compared to other risk sources.
- Bayesian or fuzzy techniques.
- Make uncertainties in risk assessment visible.
- Get quantitative results instead of conceptual/general models.
- Ways to characterize risks due to human operator error.
- Current state-of the-art paper on expert elicitation.
- Fuzzy sets.
- Value of standard basis analysis.
- Bias.

- Risk-based water management strategies and practices.
- Incorporate multi-objective analysis into risk analysis.
- Ways of presenting uncertain information to decisionmakers and the public.
- Means to make past history and development of terms, methods, and strategies available to more people in the field (especially new ones!).
- Publish more copies of proceedings so they don't go out of print so fast!
- How to arrive at acceptable level for risk management.
- Rule curves.
- Use multi-objective framework.
- Reflecting uncertainty in forecasts of demands.
- System model that incorporates the multi-objective, multivariate nature of water resource issues.
- To what extent is uncertainty information desired by decisionmakers? Do they really want it, or are we giving it to them because we think they should have it?

2000

- *Acceptable risk*
    - Honorable failure.
    - Alternatives to risk acceptance to help make decisions.
    - Can we develop a consistent integrated process to achieve acceptance of risk (e.g., public participation, measures of mitigation, roles of insurance, equity)?
    - How can we (society) come to consensus on fundamental terms of reference? I am thinking of the value of a life.
- *Risk communication*
    - Educate the public.
    - Definition between disciplines (Kaplan's "Two Rules of Communication Problems").
    - How to communicate the results of risk analysis to decisionmakers and to the attention of the public.
    - Communicating the message of risk-based decisionmaking in plain, understandable terms. These are extremely useful concepts but need to be presented in a way that can inform the decision process.
    - Develop a common vocabulary so we can talk to each other (and others) and be understood.
    - Standardization of language and terminology.
    - How do we gain broad (multi-agency) consensus for an integrated approach to infrastructure protection?
    - Understanding what decisionmakers need to make decisions.
    - Selling risk analysis.
    - Addressing events that are truly extreme; the conference topics were not really extreme.
- *Lack of information, uncertainty, and addressing extreme events*
    - Consequence management of extreme events.
    - Techniques of determining risk for extreme events.
    - Modeling extreme events (e.g., earthquake, flood, accident) and the results.
    - How to address making decisions without adequate (or any) information.
    - Tests of extreme risks based on existing techniques.
    - Means for addressing uncertainty in decisions.

- How to determine likelihood of occurrence in areas lacking statistical information (including extreme events).
- Hydraulic uncertainty.
- Uncertainty sources, types, and theories.
- Uncertainty measures and their uses in decisionmaking.
- Develop uncertainty "bounds" for waterway traffic demand, especially uncertainty associated with long-term forecasts.

- *Prioritization of multiple objectives and investment.*
  - Multiple-criteria decisionmaking in practice.
  - Decisionmaking under multi-objective tradeoff analysis.
  - Prioritizing investments to manage risks under budget constraints and tradeoffs.

- *Environment/ecosystem*
  - Use of risk analysis in ecological assessment.
  - Climate change.
  - Integrated models for ecosystem restoration.
  - Integration of environmental outputs and costs into over-all risk analysis.
  - How to deal with latent effects. This is especially relevant for environmental contamination and the eventual need to decommission/replace huge structures. Raise questions of planning and who pays and when.

- *Risk analysis applications, development of tools and paradigms*
  - Actual use of risk analysis toward decisions.
  - Risk analysis for dam safety: how we estimate probabilities, what we do with loss of life.
  - Contaminant remediation and preventing pollution of groundwater resources.
  - Application to water distribution systems.
  - More examples of actual projects and field experience.
  - Application of risk-based decisionmaking to a wider range of problems; need for scope beyond other applications.
  - Development and use of qualitative risk assessment tools.
  - Investigation and development of new paradigms for risk management.

- *Risk modeling*
  - How to assure that your "list" of failure modes is complete enough.
  - Dynamics of risk.
  - Understanding and implementing the dependencies (correlations) between random variables included in Monte Carlo analysis used for uncertainty or risk analysis.
  - How to make the formalism (mathematics) express the subjective aspect of risk.
  - Incorporating trend analysis into risk analysis.
  - Correlation among variables in analysis.
  - Nonstructured complexity.
  - Taking motivation of thinking opponents into the equation.
  - How to reduce the costs of acquiring data to perform risk analysis.
  - Expert opinion elicitation.

- *Others*
  - Further investigations into engineering risks in federal projects.
  - Presence of FEMA to discuss management strategies for flood plains.
  - Many references were made to the effect that insurance is not available to share risk. Consider a panel on risk-sharing, transfer, etc., for water projects.
  - Integrating economics into risk analysis.

2002

- How to realize an all-hazards approach.
- Where risk management fits into ecosystem restoration.
- Should we minimize regret in decisionmaking for terrorism?
- Investments in severity measures.
- Estimation of vulnerabilities.
- Risk communication among multiple stakeholders.
- How public attitude relates to the susceptibility of approaches to addressing terrorism.
- Control system vulnerability.
- How private ownership and consolidation is influencing policy decisions.
- All-hazard risk-based decisionmaking.
- Game theory for understanding terrorist risks.
- Countermeasures for cyber-terrorism.
- Threat scenarios for water resource targets.
- Means of dealing with the uncertainties in defending against terrorism.
- What advantages lie with attackers and what advantages and disadvantages lie with defenders.
- Risk-based analysis of critical infrastructures from all-hazard perspective.
- Assessment of critical infrastructure vulnerabilities that impact economies; consequences of terrorist attacks.
- Risk-based decisionmaking in a game-theory context.
- Role of predicting in risk-based decisionmaking.
- Role of regret (opportunity lost) in risk-based decisionmaking.
- Optimal allocation of resources to security.
- Applications of expected utility theory to risk-based decisionmaking.
- What to do when probabilities of threats are not well known.
- Risks to control systems.
- Vulnerabilities and risks of software system designs that govern organizations and information management.
- Better defining place and appropriate use of varying analytical tools (i.e., possibility, fuzzy math, probability).
- Explore calculi of possibility.
- Explore vulnerability of control systems.
- Explore fuzzy-logic approach.
- Explore need to change the way we do tradeoff analysis (i.e., do we fund nine projects that benefit the majority, or do we fund one expensive project that can have a significant impact on some people, if attacked?).
- More case studies to identify risk scenarios in water resources, with repository of "lessons learned."
- Considering the "state-of-the-art" in risk analysis used in various infrastructures (even other than water).
- The theme of the conference, "Risk of Terrorism," is very relevant and should be further explored in the future.
- Developing applications of risk analysis techniques to terrorists' attacks.
- Acceptance or rejection of these techniques by key decisionmakers.
- Consistency of definitions and terminology.

- Vulnerability of infrastructure systems.
- Flood evacuation for dam safety.
- Risk management of multiple hazards.
- Consequence assessment needs more study.
- Developing tools to assist risk management.
- Understanding the roles of practitioners and their interfaces with the systems they are operating, their education, their reactions to crisis.
- How is risk-based decisionmaking handled in other areas (i.e., energy, defense, law enforcement)?
- Role of complexity, adaptation, and evolution.
- How do we identify or rank threats to natural vs. man-made situations (terrorism, non-sustainable development).
- How will we accomplish risk-based decisions and management in water resources while maintaining or enhancing economic development, and with constraints such as terrorism, climate change?
- The concept or the incorporation of vulnerability of habitats, population, individuals in ecological risk assessment should be expanded. Rather than risk-based assessment on the state-of-the-system variables and the stresses, we have to look at what are the long-term threats and how vulnerable are the resources when we study the ecology.
- Risk of all-hazards and prioritization for action.
- Cyber-security.
- Risk to water resources infrastructure.
- Need international approach to this project.
- A survey tool that supports the methodologies.
- The availability of methodologies in the public domain.
- Interoperability and standardization of local/federal teams.
- The respective roles, functions, and responsibilities of various levels of government.
- Methods/models for evaluating risk/cost effectiveness of various measures designed to protect against terrorism.
- Need to make risk analysis more realistic and useful to decisionmakers.

PARTICIPANTS

UNITED ENGINEERING FOUNDATION CONFERENCE
RISK-BASED DECISIONMAKING IN WATER RESOURCES X

November 3 - 8, 2003
Santa Barbara, California

BIER, VICKI
    University of Wisconsin-Madison

BOGARDI, ISTVAN
    University of Nebraska

BOWLES, DAVID
    Utah State University

CONNER, RONALD
    US Army Corps of Engineers

CROWTHER, KENNETH
    University of Virginia

DICDICAN, RUTH
    University of Virginia

EZELL, BARRY
    Old Dominion University

FERSON, SCOTT
    Applied Biomathematics

HAIMES, YACOV
    University of Virginia

HECKER, ED
    US Army Corps of Engineers

KAPLAN, STANLEY
    Food Safety and Inspection Service, US Department of Agriculture

KAUTH, DAVID
    US Air Force

KELLEY, RICHARD
    University of Iowa

LAMBERT, JAMES H.
    University of Virginia

LONGSTAFF, THOMAS
    Carnegie Mellon University – Softwear Engineering Institute

LOUCKS, DANIEL P.
    Cornell University

MATALAS, NICHOLAS
    Hydrologist

MOSER, DAVID
    US Army Corps of Engineers

NANDA, S.K.
    US Army Corps of Engineers

O'CONNOR, ROBERT
    National Science Foundation

PATEV, ROBERT
    US Army Corps of Engineers

PODOLSKE, LEWIS
    Office of Homeland Security

ROWE, WILLIAM
    Rowe Research & Engineering Associates, Inc.

RUIZ, CARLOS
    US Army Corps of Engineers

SANTOS, JOOST
    University of Virginia

SAUNDERS, JOHN
    National Defense University

STAKHIV, EUGENE
    US Army Corps of Engineers

WEISS, JOSEPH
    KEMA Consulting

WILLARD, DAN
    Indiana University

RISK-BASED DECISIONMAKING
IN WATER RESOURCES
X

Radisson Santa Barbara Hotel
Santa Barbara, California

November 3 - 8, 2002

Conference Chair: Yacov Y. Haimes
University of Virginia

Conference Co-Chairs: David A. Moser, Eugene Stakhiv
US Army Corps of Engineers

Conference Liaison: Richard Fein

Local Coordinator: Antoinette Chartier

Sponsor: United Engineering Foundation, Inc.
Three Park Ave, 27th Floor
New York, NY 10016-5902

Proceedings Editors: Yacov Y. Haimes, David A. Moser, Eugene Z. Stakhiv
Technical Editor: Grace Ivry Zisk
Technical Associates: Della Dirickson, Burton I. Zisk

All meetings held in the *El Cabrillo Room* and all meals in *La Cantina*.

Sunday, November 3, 2002

17:30—18:30   CONFERENCE REGISTRATION

18:30—19:30   DINNER AND OPENING REMARKS—*La Cantina*

Monday, November 4, 2002

07:00—08:00   BREAKFAST BUFFET

08:15—08:30   **WELCOME REMARKS**—*El Cabrillo*

Yacov Haimes, Conference Chair
David Moser, Conference Co-chair
Dick Fein, UEF Conference Liaison

08:30—12:00   **SESSION 1—FIRST PLENARY SESSION**
Risks of Terrorism to the Homeland

Chair: Yacov Y. Haimes, University of Virginia
Rapporteurs: Kenneth Crowther, Bill Rowe

- **Yacov Y. Haimes**, University of Virginia
- **Thomas Longstaff**, Carnegie Mellon—Software Engineering Institute
- **Robert O'Connor**, Pennsylvania State University/NSF
- **Joe Weiss**, KEMA Consulting
- **Dan Willard**, Indiana University

Review of governmental activities starting with the President's Commission on Critical Infrastructures Protection (PCCIP) and the risks of terrorism to the homeland, focusing on the water resources system of systems.

10:00—10:30   COFFEE BREAK

10:30   **SESSION 1 RESUMES**

12:00—13:00   LUNCH

14:00—17:00   AD HOC SESSIONS and/or FREE TIME

17:00—18:00   SOCIAL HOUR—*La Cantina*

18:00—19:00   DINNER

19:00—21:30   SESSION 2—The Vulnerabilities of the Homeland's Water
                 Resources System of Systems

                 Chair: S.K. Nanda, US Army Corps of Engineers
                 Rapporteur: Barry Ezell

- Yacov Haimes, Jim Lambert, University of Virginia
  Reducing Vulnerability of Water Supply Systems to Attack
- Joe Weiss, KEMA Consulting
  Technical Status of Cyber Security of Process Control Systems
- Dan Willard, Indiana University—Restoring the Everglades by Reducing the Potential Risk of Human Interference
- Bob Patev, US Army Corps of Engineers—Design and Planning of Infrastructure Projects Subjected to Extreme Events

The extent to which the Homeland's water resources system of systems is vulnerable to terrorism.

21:30—22:15 SOCIAL HOUR

Tuesday, November 5, 2002

07:00—08:00   BREAKFAST BUFFET

08:00—12:00   SESSION 3—Lessons Learned from Experience Dealing with
                 Risks of Extreme Events: Part I

                 Chair: Jim Lambert, University of Virginia
                 Rapporteur: Ruth Dicdican

- Jim Lambert, University of Virginia—Infrastructure Terrorism Risks Characterized by the Superposition of Networks
- David Bowles, Utah State University
  Extreme Events in Dam Safety Risk Assessment
- Scott Ferson, Applied Biomathematics
  Terrorism's Implications for Uncertainty Calculi
- Istvan Bogardi, University of Nebraska
- The Probabilistic versus Fuzzy Logic Paradigm in Risk Analysis
- Thomas Longstaff, Carnegie Mellon – Software Engineering Institute
  Information Assurance and Cyber Security

What can the professional community learn from our past experience in dam safety, flood warning and evacuation, earthquake studies, and the management of risks of extreme events?

10:00—10:30   COFFEE BREAK

10:30   SESSION 3 RESUMES

12:00—13:00   LUNCH

14:00—17:00   AD HOC SESSIONS and/or FREE TIME

17:00—18:00   SOCIAL HOUR

18:00—19:00   DINNER

19:00—21:30   **SESSION 4—Lessons Learned from Experience Dealing with Risks of Extreme Events: Part II**

Chair: Robert O'Connor: National Science Foundation
Rapporteur: Joost Santos

- **John Saunders,** National Defense University—**A Dynamic Risk Model for Information Technology Security in a Critical Infrastructure Environment**
- **Pete Loucks,** Cornell University—**Quantifying and Communicating Model Uncertainty for Decisionmaking in the Florida Everglades**
- **Vicki Bier,** University of Wisconsin-Madison
  Game-theoretic Models for Critical Infrastructure Protection
- **Stan Kaplan,** Food Safety and Inspection Service, USDA
  On the Application of the General Theory of Quantitative Risk Assessment (GTQRA) to Combating Terrorism
- **Nick Matalas,** Hydrologist—**Perspectives on the Characteristics of the Vulnerability of Water Resource Systems to Nature and Terror**

Extreme and catastrophic events are often underestimated and commensurated in the risk assessment and management process with other more likely or less consequential events. Identify theory and methodology that can be adapted to address the risks of terrorism to the Homeland's water resources system of systems.

21:30—22:15   SOCIAL HOUR

Wednesday, November 6, 2002

07:00—08:00   BREAKFAST

08:00—12:00   SESSION 5—The Interconnectedness and Interdependencies between the Water Resources System of Systems and Other Infrastructures

               Chair: Istvan Bogardi, University of Nebraska
               Rapporteur: Ron Conner, USACE

- **Barry Ezell,** Old Dominion University
  Vulnerability Assessment of Critical Infrastructures
- **Joost Santos,** University of Virginia—**Demand-Reduction Input-Output (I-O) Inoperability Modeling of Infrastructure Interconnectedness**
- **Robert O'Connor,** Pennsylvania State University/NSF—**Belief Systems and Reducing Risks from Terrorism**
- **Dan Willard,** Indiana University—**Critical Infrastructures and Ecology**

How to model, assess, and manage the risks of terrorism to the homeland's water resources system of systems, focusing on the interconnectedness and interdependencies with other infrastructures.

10:00—10:30   COFFEE BREAK

10:30   **SESSION 5 RESUMES**

12:00—13:00   LUNCH

14:00—17:00   AD HOC SESSIONS and/or FREE TIME

17:00—18:00   SOCIAL HOUR

18:00—19:00   DINNER

19:00—21:30   **SESSION 6—Perspectives of Infrastructure Survivability**

               Chair: Dave Moser: US Army Corps of Engineers
               Rapporteur: John Saunders

- **Ron Conner,** US Army Corps of Engineers
  Economic Consequences and Disaster Response

- **Lew Podolske,** Office of Homeland Security
  Overview of Economic Consequences Policy Coordinating Committee on Economic Recovery Strategy for Large-Scale Disasters

- **David Bowles,** Utah State University
  GIS Model for Predicting Dam Failure Consequences

Explore risk-benefit-cost justifications for the assurance of survivability of critical water resource infrastructures.

21:30—22:15   SOCIAL HOUR

Thursday, November 7, 2002

07:00—08:00   BREAKFAST BUFFET

08:00—12:00   **SESSION 7—Institutional and Organizational Restructuring**

   Chair: **Gene Stakhiv,** US Army Corps of Engineers
   Rapporteur: Scott Ferson

- **Lew Podolske**, Office of Homeland Security
  Restructuring for Emergency Response to Large-Scale Disasters
- **Bill Rowe,** Rowe Research & Engineering Associates, Inc.
  Vulnerability to Terrorism: Addressing the Human Variables
- **Ed Hecker,** US Army Corps of Engineers
  Restructuring for Emergency Response to Large-Scale Disasters
- **Yacov Haimes,** University of Virginia
  The Role of Modeling in the Assessment and Strategic Management of Risks of Terrorism to Our Homeland

Are current institutions and organizations at the federal, state, and local levels adequate to address the risks of terrorism to the Homeland's water resources system of systems? If not, what should be done?

10:00—10:30   COFFEE BREAK

10:30   **SESSION 7 RESUMES**

12:00—13:00   LUNCH
14:00—17:00   AD HOC SESSIONS and/or FREE TIME
17:00—18:00   SOCIAL HOUR

18:00—19:00   CONFERENCE CLOSING DINNER

19:00—21:30   **SESSION 8—Panel Discussion: Synthesis—
              What Does It All Mean?**

              Chair: Yacov Y. Haimes: University of Virginia
              Rapporteur: Jim Lambert

- **Nick Matalas,** Hydrologist
- **Bill Rowe,** QURENT Systems, Inc.
- **Gene Stakhiv,** US Army Corps of Engineers
- **Stan Kaplan,** Food Safety and Inspection Service, USDA
- **Vicki Bier,** University of Wisconsin-Madison
- **Scott Ferson,** US Army Corps of Engineers
- **Yacov Haimes,** University of Virginia

21:30—22:15   SOCIAL HOUR

Friday, November 8, 2002

07:00—08:15   BREAKFAST BUFFET

08:30—9:30   SUMMARY and CONCLUSION OF CONFERENCE

Subject Index

Page number refers to the first page of paper

California, 146
Colorado River, 146
Communication, 40

Dam failure, 126
Dam safety, 126
Decision making, 40
Disasters, 1, 59, 91
Dynamic models, 23, 163

Earthquakes, 1
Emergency services, 126, 160

Failure modes, 1
Federal government, 82, 146
Flood damage, 12
Floods, 1
Florida, 40
Fuzzy sets, 12

Geographic information systems, 126

Hurricanes, 1, 160

Information technology (IT), 23, 163
Infrastructure, 1, 23, 91, 104, 155, 163, 168, 171

Methodology, 91
Models, 23, 40, 91, 126

Optimization, 59

Restoration, 40
Risk analysis, 1, 12, 23, 77, 91, 119, 155, 163
Risk management, 59, 104, 126, 160, 163, 168, 171

Security, 23, 59, 104, 163, 168, 171

Terrorism, 1, 59, 77, 82, 91, 119, 155

Uncertainty analysis, 12, 40, 160

Water policy, 146
Water resources, 82, 119
Water resources management, 146
Water supply, 91

Author Index

Page number refers to the first page of paper

Abhichandani, Vinod, 59
Aboelata, Maged, 126

Bier, Vicki M., 59
Bogardi, Istvan, 12
Bowles, David S., 126

Dicdican, Ruth Y., 160
Duckstein, Lucien, 12

Ezell, Barry C., 91

Haimes, Yacov Y., 104, 171

Kaplan, Stan, 77

Lambert, James H., 1
Lambert, Jim, 168
Loucks, Daniel P., 40

Matalas, Nicholas C., 82
McClelland, Duane M., 126

OÕConnor, Robert E., 119

Rowe, William D., 155

Santos, Joost, 163
Santos, Joost R., 104
Sarda, Priya, 1
Saunders, John H., 23
Stakhiv, Eugene Z., 146